Mary Noailles Murfree

The Phantoms of the Foot-Bridge

And Other Stories

Mary Noailles Murfree

The Phantoms of the Foot-Bridge
And Other Stories

ISBN/EAN: 9783337004798

Printed in Europe, USA, Canada, Australia, Japan

Cover: Foto ©berggeist007 / pixelio.de

More available books at **www.hansebooks.com**

CONTENTS

ILLUSTRATIONS

THE PHANTOMS OF THE FOOT-BRIDGE

Across the narrow gorge the little foot-bridge stretched—a brace of logs, the upper surface hewn, and a slight hand-rail formed of a cedar pole. A flimsy structure, one might think, looking down at the dark and rocky depths beneath, through which flowed the mountain stream, swift and strong, but it was doubtless substantial enough for all ordinary usage, and certainly sufficient for the imponderable and elusive travellers who by common report frequented it.

"We ain't likely ter meet nobody. Few folks kem this way nowadays, 'thout it air jes' ter ford the creek down along hyar a piece, sence harnts an' sech onlikely critters hev been viewed a-crossin' the foot-bredge. An' it hev got the name o' bein' toler'ble onlucky, too," said Roxby.

His interlocutor drew back slightly. He had his own reasons to recoil from the subject of death. For him it was invested with a more immediate terror than is usual to many of the living, with that flattering persuasion of immortality in every strong pulsation repudiating all possibility of cessation. Then, lifting his gloomy, long-lashed eyes to the bridge far up the stream, he asked, "Whose 'harnts'?"

His voice had a low, repressed cadence, as of one who speaks seldom, grave, even melancholy, and little indicative of the averse interest that had kindled in his sombre eyes. In comparison the drawl of the mountaineer, who had found him heavy company by the way, seemed imbued with an abnormal vivacity, and keyed a tone or two higher than was its wont.

"Thar ain't a few," he replied, with a sudden glow of the pride of the cicerone. "Thar's a grave-yard t'other side o' the gorge, an' not more than a haffen-mile off, an' a cornsider'ble passel o' folks hev been buried thar off an' on, an' the foot-bredge ain't in nowise ill-convenient·ter them."

Thus demonstrating the spectral resources of the locality, he rode his horse well into the stream as he spoke, and dropped the reins that the animal's impatient lips might reach the water. He sat facing the foot-bridge, flecked with the alternate shifting of the sunshine and the shadows of the tremulous firs that grew on either side of the high banks on the ever-ascending slope, thus arching both above and below the haunted bridge. His companion had joined him in the centre of the stream ; but while the horses drank, the stranger's eyes were persistently bent on the concentric circles of the water that the movement of the animals had set astir in the current, as if he feared that too close or curious a gaze might discern some pilgrim, whom he cared not to see, traversing that shadowy quivering foot-bridge. He was mounted on a strong, handsome chestnut, as marked a contrast to his guide's lank and trace-galled sorrel as were the two riders. A slender gloved hand had fallen with the reins to the pommel of the saddle. His soft felt hat, like a sombrero, shadowed his clear-cut face. He was carefully shaven, save for a long drooping dark mustache and imperial. His suit of dark cloth was much concealed by a black cloak, one end of which thrown back across his shoulder showed a bright blue lining, the color giving a sudden heightening touch to his attire, as if he were "in costume." It was a fleeting fashion of the day, but it added a certain picturesqueness to a horseman, and seemed far enough from the times that produced the square-tailed frock-coat which the mountaineer wore, constructed of brown jeans, the skirts of which stood stiffly out on each side of the saddle, and gave him, with his broad-brimmed hat, a certain Quakerish aspect.

"I dun'no' why folks be so 'feared of 'em," Roxby remarked, speculatively. "The dead ain't so oncommon, nohow. Them ez hev been in the war, like you an' me done, oughter be in an' 'bout used ter corpses—though I never seen none o' 'em afoot agin. Lookin' at a smit field o' battle, arter the rage is jes' passed, oughter gin a body a realizin' sense how easy the sperit kin flee, an' what pore vessels fur holdin' the spark o' life human clay be."

Simeon Roxby had a keen, not unkindly face, and he had that look of extreme intelligence which is entirely distinct from intellectuality, and which one sometimes sees in a minor degree in a very clever dog or a fine horse. One might rely on him to understand instinctively everything one might say to him, even in its subtler æsthetic values, although he had consciously learned little. He was of the endowed natures to whom much is given, rather than of those who are set to acquire. He had many lines in his face—even his simple life had gone hard with him, its sorrows unassuaged by its simplicity. His hair was grizzled, and hung long and straight on his collar. He wore a grizzled beard cut broad and short. His boots had big spurs, although the lank old sorrel had never felt them. He sat his horse like the cavalryman he had been for four years of hard riding and raiding, but his face had a certain gentleness that accented the Quaker-like suggestion of his garb, a look of communing with the higher things.

" I never blamed 'em," he went on, evidently reverting to the spectres of the bridge—" I never blamed 'em for comin' back wunst in a while. It 'pears ter me 'twould take me a long time ter git familiar with heaven, an' sociable with them ez hev gone before. An', my Lord, jes' think what the good green yearth is! Leastwise the mountings. I ain't settin' store on the valley lands I seen whenst I went ter the wars. I kin remember yit what them streets in the valley towns smelt like."

He lifted his head, drawing a long breath to inhale the exquisite fragrance of the fir, the freshness of the pellucid water, the aroma of the autumn wind, blowing through the sere leaves still clinging red and yellow to the boughs of the forest.

" Naw, I ain't blamin' 'em, though I don't hanker ter view 'em," he resumed. " One of 'em I wouldn't be afeard of, though. I feel mighty sorry fur her. The old folks used ter tell about her. A young 'oman she war, a-crossin' this bredge with her child in her arms. She war young, an' mus' have been keerless, I reckon; though ez 'twar her fust baby, she moughtn't hev been practised in holdin' it an' sech, an' somehows it slipped through her arms an' fell inter the ruver, an' war killed in a minit, dashin' agin the rocks. She jes' stood fur a second a-screamin' like a wild painter, an' jumped off'n the bredge arter it. She got it agin; for when they dragged her body out'n the ruver she hed it in her arms too tight fur even death ter onloose. An' thar they air together in the buryin'-ground."

He gave a nod toward the slope of the mountain that intercepted the melancholy view of the grave-yard.

"Got it yit!" he continued; ", bekase" (he lowered his voice) "on windy nights, whenst the moon is on the wane, she is viewed kerryin' the baby along the bredge — kerryin' it clear over, *safe an' sound*, like she thought she oughter done, I reckon, in that one minute, whilst she stood an' screamed an' surveyed what she hed done. That child would hev been nigh ter my age ef he hed lived."

Only the sunbeams wavered athwart the bridge now as the firs swayed above, giving glimpses of the sky, and their fibrous shadows flickered back and forth. The wild mountain stream flashed white between the brown bowlders, and plunged down the gorge in a succession of cascades, each seeming more transparently green and amber and brown than the other. The chestnut horse gazed meditatively at these limpid out-gushings, having drunk his fill; then thought better of his moderation, and once

more thrust his head down to the water. The hand
of his rider, which had made a motion to gather up
the reins, dropped leniently on his neck, as Simeon
Roxby spoke again :

"Several—several others hev been viewed, actin'
accordin' ter thar motions in life. Now thar war a
peddler—some say he slipped one icy evenin', 'bout
dusk in winter — some say evil ones waylaid him
fur his gear an' his goods in his pack, but the settlemint mostly believes he war alone whenst he fell.
His pack 'pears ter be full still, they say — but ye
air 'bleeged ter know he hev hed ter set that pack
down fur good 'fore this time. We kin take nuthin'
out'n this world, no matter what kind o' a line o'
goods we kerry in life. Heaven's no place fur
tradin', I understan', an' I *do* wonder sometimes
how in the worl' them merchants an' sech in the
valley towns air goin' ter entertain tharse'fs in the
happy land o' Canaan. It's goin' ter be sorter bleak
fur them, sure's ye air born."

With a look of freshened recollection, he suddenly drew a plug of tobacco from his pocket, and
he talked on even as he gnawed a piece from it.

"Durin' the war a cavalry-man got shot out hyar
whilst runnin' 'crost that thar foot-bredge. Thar
hed been a scrimmage an' his horse war kilt, an'
he tuk ter the bresh on foot, hopin' ter hide in the
laurel. But ez he war crossin' the foot-bredge some
o' the pursuin' party war fordin' the ruver over
thar, an' thinkin' he'd make out ter escape they
fired on him, jes' ez the feller tried ter surrender.
He turned this way an' flung up both arms—but
thar's mighty leetle truce in a pistol-ball. That

minute it tuk him right through the brain. Seems
toler'ble long range fur a pistol, don't it? He kin
be viewed now most enny moonlight night out hyar
on the foot-bredge, throwin' up both hands in sign
of surrender."

The wild-geese were a-wing on the way southward.
Looking up to that narrow section of the blue sky
which the incision of the gorge into the very depths
of the woods made visible, he could see the tiny
files deploying along the azure or the flecking cir-
rus, and hear the vague clangor of their leader's cry.
He lifted his head to mechanically follow their
flight. Then, as his eyes came back to earth, they
rested again on the old bridge.

"Strange enough," he said, suddenly, "the sker-
riest tale I hev ever hearn 'bout that thar old bredge
is one that my niece set a-goin'. She *seen* the harnt
herself, an' it shakes me wuss 'n the idee o' all the
rest."

His companion's gloomy gaze was lifted for a
moment with an expression of inquiry from the
slowly widening circles of the water about the
horse's head as he drank. But Roxby's eyes, with
a certain gleam of excitement, a superstitious di-
lation, still dwelt upon the bridge at the end of
the upward vista. He went on merely from the
impetus of the subject. "Yes, sir — she *seen* it a-
pacin' of its sorrowful way acrost that bredge, same
ez the t'others of the percession o' harnts. 'Twar
my niece, Mill'cent — brother's darter — by name,
Mill'cent Roxby. Waal, Mill'cent an' a lot o'
young fools o' her age—little over fryin' size—they
'tended camp-meetin' down hyar on Tomahawk

Creek — 'tain't so long ago — along with the old
folks. An' 'bout twenty went huddled up tergether
in a road-wagin. An', lo! the wagin it bruk down
on the way home, an' what with proppin' it up on a
crotch, they made out ter reach the cross-roads
over yander at the Notch, an' thar the sober old
folks called a halt, an' hed the wagin mended at
the blacksmith-shop. Waal, it tuk some two hours,
fur Pete Rodd ain't a-goin' ter hurry hisself — in
my opinion the angel Gabriel will hev ter blow his
bugle oftener'n wunst at the last day 'fore Pete
Rodd makes up his mind ter rise from the dead
an' answer the roll-call — an' this hyar young lot
sorter found it tiresome waitin' on thar elders'
solemn company. The old folks, whilst waitin',
set outside on the porches of the houses at the
settlemint, an' repeated some o' the sermons they
hed hearn at camp, an' more'n one raised a hyme
chune. An' the young fry—they hed hed a steady
diet o' sermons an' hyme chunes fur fower days—
they tuk ter stragglin' off down the road, two an'
two, like the same sorter idjits the world over,
leavin' word with the old folks that the wagin
would overtake 'em an' pick 'em up on the road
when it passed. Waal, they walked several mile,
an' time they got ter the crest o' the hill over
yander the moon hed riz, an' they could look down
an' see the mist in the valley. The moon war
bright in the buryin'-groun' when they passed it,
an' the head-boards stood up white an' stiff, an' a
light frost hed fell on the mounds, an' they showed
plain, an' shone sorter lonesome an' cold. The
young folks begun ter look behind em' fur the

wagin. Some said—I b'lieve 'twar Em'ry Keenan
—they could read the names on the boards plain,
'twar so light, the moon bein' nigh the full: but
Em'ry never read nuthin' at night by the moon in
his life; he ain't enny too capable o' wrastlin' with
the alphabet with a strong daytime on his book ter
light him ter knowledge. An' the shadows war
black an' still, an' all the yearth looked ez ef
nuthin' lived nor ever would agin, an' they hearn a
wolf howl. Waal, that disaccommodated the gals
mightily, an' they hed a heap more interes' in that
old wagin, all smellin' rank with wagin-grease an'
tar, than they did in thar lovyers; an' they hed
ruther hev hearn that old botch of a wheel that Pete
Rodd hed set onto it comin' a-creakin' an' a-com-
plainin' along the road than the sweetest words
them boys war able ter make up or remember. So
they stood thar in the road — a-stare-gazin' them
head-boards, like they expected every grave ter
open an' the reveilly ter sound — a-waitin' ter be
overtook by the wagin, a-listenin', but hearin' nuthin' in the silence o' the frost — not a dead leaf a-
twirlin', nor a frozen blade o' grass astir. An' then
two or three o' the gals 'lowed they hed ruther
walk back ter meet the wagin, an' whenst the boys
'lowed ter go on—nuthin' war likely ter ketch 'em
—one of 'em bust out a-cryin'. Waal, thar war the
eend o' that much! So the gay party set out on
the back track, a-keepin' step ter sobs an' sniffles,
an' that's how kem *they* seen no harnt. But Mill'-
cent an' three or four o' the t'others 'lowed they'd
go on. They warn't two mile from home, an' full
five from the cross-roads. So Em'ry Keenan — he

hev been waitin' on her sence the year one—so he
put his skeer in his pocket an' kem along with her,
a-shakin' in his shoes, I'll be bound! So down the
hill in the frosty moonlight them few kem — purty
nigh beat out, I reckon, Mill'cent war, what with
the sermonizin' an' the hyme-singin' an' hevin' ter
look continual at the sheep's-eyes o' Em'ry Keenan
—he wears my patience ter the bone! So she concluded ter take the short-cut. An' Em'ry he
agreed. So they tuk the lead, the rest a followin',
an' kem down thar through all that black growth"—
he lifted his arm and pointed at the great slope,
dense with fir and pine and the heavy underbrush—
"keepin' the bridle-path—easy enough even at night,
fur the bresh is so thick they couldn't lose thar way.
But the moonlight war mightily slivered up, fallin'
through the needles of the pines an' the skeins of
dead vines, an' looked bleached and onnatural, an'
holped the dark mighty leetle. An' they seen the
water a-shinin' an' a-plungin' down the gorge, an'
the glistenin' of the frost on the floor o' the bredge.
Thar war a few icicles on the hand-rail, an' the
branches o' the firs hung ez still ez death; only
that cold, racin', shoutin', jouncin' water moved.
Jes ez they got toler'ble nigh the foot-bredge a
sudden cloud kem over the face o' the sky. Thar
warn't no wind on the yearth, but up above the
air war a-stirrin'. An' Em'ry he 'lowed Mill'cent
shouldn't cross the foot-bredge whilst the light
warn't clar — I wonder the critter hed that much
sense! An' she jes' drapped down on that rock
thar ter rest"—he pointed up the slope to a great
fragment that had broken off from the ledges and

THE PHANTOM OF THE FOOT-BRIDGE

lay near the bank : the bulk of the mass wás overgrown with moss and lichen, but the jagged edges
of the recent fracture gleamed white and crystalline
among the brown and olive-green shadows about
it.   A tree was close beside it.   "Agin that thar
pine trunk Em'ry he stood an' leaned.   The rest
war behind, a-comin' down the hill.   An' all of a
suddenty a light fell on the furder eend o' the foot-
bredge—a waverin' light, mighty white an' misty in
the darksomeness.   Mill'cent 'lowed ez fust she
thunk it war the moon.   An' lookin' up, she seen
the cloud ; it held the moon close kivered.   An'
lookin' down, she seen the light war· movin'—
movin' from the furder eend o' the bredge, straight
acrost it.   Sometimes a hand war held afore it, ez
ef ter shield it from the draught, an' then Mill'cent
'seen twar a candle, an' the white in the mistiness
war a 'oman wearin' white an' carryin' it.   Lookin'
ter right an' then ter lef' the 'oman kem, with now
her right hand shieldin' the candle she held, an'
now layin' it on the hand-rail.   The candle shone
on the water, fur it didn't flare, an' when the 'oman
held her hand before it the light made a bright
spot on the foot-bredge an' in the dark air about
her, an' on the fir branches over her head.   An' a
thin mist seemed to hang about her white frock,
but not over her face, fur when she reached the
middle o' the foot-bredge she laid her hand agin
on the rail, an' in the clear light o' the candle
Mill'cent seen the harnt's face.   An' thar she beheld her own face ; *her own face* she looked upon
ez she waited thar under the tree watchin' the foot-
bredge ; *her own face* pale an' troubled ; her own

self dressed in white, crossin' the foot-bredge, an' lightin' her steps with a corpse's candle." He drew up the reins abruptly. He seemed in sudden haste to go.

His companion looked with deepening interest at the bridge, although he followed his guide's surging pathway to the opposite bank. As the two dripping horses struggled up the steep incline he asked, "Did the man with her see the manifestation also?"

"He '*lows* he did," responded Roxby, equivocally. "But when Mill'cent fust got so she could tell it, 'peared ter me ez Em'ry Keenan fund it ez much news ez the rest o' we-uns. Mill'cent jes' drapped stone-dead, accordin' ter all accounts, an' he an' the t'other young folks flung water in her face till she kem out'n her faint; an' jes' then they hearn the wagin a-rattlin' along the road, an' they stopped it an' fetched her home in it. She never told the tale till she war home, an' it skeered me an' my mother powerful, fur Mill'cent is all the kin we hev got. Mill'cent is gran'daddy an' gran'mammy, sons an' daughters, uncles an' aunts, cousins, nieces, an' nephews, all in one. The only thing I ain't pervided with is a nephew-in-law, an' I don't need him. Leastwise I ain't lookin' fur Em'ry Keenan jes' at present."

The pace was brisker when the two horses, bending their strength sturdily to the task, had pressed up the massive slope from the deep cleft of the gorge. As the road curved about the outer verge of the mountain, the valley far beneath came into view, with intersecting valleys and transverse ranges, dense with the growths of primeval wilder-

nesses, and rugged with the tilted strata of great
upheavals, and with chasms cut in the solid rock
by centuries of erosion, traces of some remote cata-
clysmal period, registering thus its throes and turmoils.   The blue sky, seen beyond a gaunt profile
of one of the farther summits that defined its
craggy serrated edge against the ultimate distances
of the western heavens, seemed of a singularly
suave tint, incongruous with the savagery of the
scene, which clouds and portents of storm might
better have befitted.   The little graveyard, which
John Dundas discerned with recognizing eyes, albeit they had never before rested upon it, was revealed suddenly, lying high on the opposite side of
the gorge.   No frost glimmered now on the lowly
mounds; the flickering autumnal sunshine loitered
unafraid among them, according to its languid
wont for many a year.   Shadows of the gray un-
painted head-boards lay on the withered grass,
brown and crisp, with never a cicada left to break
the deathlike silence.   A tuft of red leaves, vagrant
in the wind, had been caught on one of the primitive monuments, and swayed there with a decora-
tive effect.   The enclosure seemed, to unaccustomed eyes, of small compass, and few the denizens
who had found shelter here and a resting-place, but
it numbered all the dead of the country-side for
many a mile and many a year, and somehow the
loneliness was assuaged to a degree by the reflection that they had known each other in life, unlike
the great herds of cities, and that it was a common
fate which the neighbors, huddled together, encountered in company.

It had no discordant effect in the pervasive sense
of gloom, of mighty antagonistic forces with which
the scene was replete; it fostered a realization of
the pitiable minuteness and helplessness of human
nature in the midst of the vastness of inanimate
nature and the evidences of infinite lengths of forgotten time, of the long reaches of unimagined history, eventful, fateful, which the landscape at once
suggested and revealed and concealed.

Like the sudden flippant clatter of castanets in
the pause of some solemn funeral music was the
impression given by the first glimpse along the
winding woodland way of a great flimsy white
building, with its many pillars, its piazzas, its "ob-
servatory," its band-stand, its garish intimations
of the giddy, gay world of a summer hotel.   But,
alack! it, too, had its surfeit of woe.

"The guerrillas an' bushwhackers tuk it out on
the old hotel, sure!" observed Sim Roxby, by way
of introduction.   "Thar warn't much fightin' hyar-
abouts, an' few sure-enough soldiers ever kem
along.   But wunst in a while a band o' guerrillas
went through like a suddint wind-storm, an' I tell
ye they made things whurl while they war about it.
They made a sorter barracks o' the old place.
Looks some like lightning hed struck it."

He had reined up his horse about one hundred
yards in front of the edifice, where the weed-grown
gravelled drive—carefully tended ten years agone
—had diverged from the straight avenue of pop-
lars, sweeping in a circle around to the broad flight
of steps.

"Though," he qualified abruptly, as if a sudden

thought had struck him, "ef ye air countin' on buyin' it, a leetle money spent ter keerful purpose will go a long way toward makin' it ez good ez new."

His companion did not reply, and for the first time Roxby cast upon him a covert glance charged with the curiosity which would have been earlier and more easily aroused in another man by the manner of the stranger. A letter—infrequent missive in his experience—had come from an ancient companion-in-arms, his former colonel, requesting him in behalf of a friend of the old commander to repair to the railway station, thirty miles distant, to meet and guide this prospective purchaser of the old hotel to the site of the property. And now as Roxby looked at him the suspicion which his kind heart had not been quick to entertain was seized upon by his alert brain.

"The cunnel's been fooled somehows," he said to himself.

For the look with which John Dundas contemplated the place was not the gaze of him concerned with possible investment—with the problems of repair, the details of the glazier and the painter and the plasterer. The mind was evidently neither braced for resistance nor resigned to despair, as behooves one smitten by the foreknowledge of the certainty of the excess of the expenditures over the estimates. Only with pensive, listless melancholy, void of any intention, his eyes traversed the long rows of open doors, riven by rude hands from their locks, swinging helplessly to and fro in the wind, and giving to the deserted and desolate old place a

spurious air of motion and life. Many of the shutters had been wrenched from their hinges, and
lay rotting on the floors. The ball-room windows
caught on their shattered glass the reflection of the
clouds, and it seemed as if here and there a wan
face looked through at the riders wending along the
weed-grown path. Where so many faces had been
what wonder that a similitude should linger in the
loneliness ! The pallid face seemed to draw back as
they glanced up while slowly pacing around the
drive. A rabbit sitting motionless on the front piazza did not draw back, although observing them with
sedate eyes as he poised himself upright on his
haunches, with his listless fore-paws suspended in
the air, and it occurred to Dundas that he was
probably unfamiliar with the presence of human
beings, and had never heard the crack of a gun. A
great swirl of swallows came soaring out of the big
kitchen chimneys and circled in the sky, darting
down again and again upward. Through an open
passage was a glimpse of a quadrangle, with its
weed-grown spaces and litter of yellow leaves. A
tawny streak, a red fox, sped through it as Dundas
looked. A half-moon, all a-tilt, hung above it. He
saw the glimmer through the bare boughs of the
leafless locust-trees here and there still standing,
although outside on the lawn many a stump bore
token how ruthlessly the bushwhackers had furnished their fires.

"That thar moon's a-hangin' fur rain," said the
mountaineer, commenting upon the aspect of the lu-
minary, which he, too, had noticed as they passed.
"I ain't s'prised none ef we hev fallin' weather

agin 'fore day, an' the man—by name Morgan
Holden — that hev charge o' the hotel property
can't git back fur a week an' better."

A vague wonder to find himself so suspicious
flitted through his mind, with the thought that
perhaps the colonel might have reckoned on this
delay. "Surely the ruvers down yander at Knox-
ville mus' be a-boomin', with all this wet weather,"
he said to himself.

Then aloud : "Morgan Holden he went ter Col-
bury ter 'tend ter some business in court, an' the
ruvers hev riz so that, what with the bredges bein'
washed away an' the fords so onsartain an' tricky,
he'll stay till the ruver falls. He don't know ye
war kemin', ye see. The mail-rider hev quit, 'count
o' the rise in the ruver, an' thar's no way ter git
word ter him. Still, ef ye air minded ter wait, I'll
be powerful obligated fur yer comp'ny down ter my
house till the ruver falls an' Holden he gits back."

The stranger murmured his obligations, but his
eyes dwelt lingeringly upon the old hotel, with its
flapping doors and its shattered windows. Through
the recurrent vistas of these, placed opposite in the
rooms, came again broken glimpses of the grassy
space within the quadrangle, with its leafless locust-
trees, first of all to yield their foliage to the autumn
wind, where a tiny owl was shrilling stridulously
under the lonely red sky and the melancholy moon.

"Hed ye 'lowed ter put up at the old hotel ?"
asked Roxby, some inherent quickness supplying
the lack of a definite answer.

For the first time the stranger turned upon him
a look more expressive than the casual fragmen-

tary attention with which he had half heeded, half
ignored his talk since their first encounter at the
railway station.

"A simple fellow, but good as gold," was the
phrase with which Simeon Roxby had been com-
mended as guide and in some sort guard.

"Not so simple, perhaps," the sophisticated
man thought as their eyes met. Not so simple but
that the truth must serve. "The colonel suggested
that it might be best," he replied, more alert to the
present moment than his languid preoccupation
had heretofore permitted.

The answer was good as far as it went. A few
days spent in the old hostelry certainly would serve
well to acquaint the prospective purchaser with
its actual condition and the measures and means
needed for its repair; but as Sim Roxby stood
there, with the cry of the owl shrilling in the desert air, the lonely red sky, the ominous tilted
moon, the doors drearily flapping to and fro as the
wind stole into the forlorn and empty place and
sped back affrighted, he marvelled at the refuge
contemplated.

"I believe there is some of the furniture here
yet. We could contrive to set up a bed from what
is left. The colonel could make it all right with
Holden, and I could stay a day or two, as we origi-
nally planned."

"Ye-es. I don't mind Holden: a man ain't
much in charge of a place ez ain't got a lock or a
key ter bless itself with, an' takes the owel an' the
fox an' the gopher fur boarders; but, ennyhow,
kem with me home ter supper. Mill'cent will hev

it ready by now ennyhows, an' ye need suthin'
hearty an' hot ter stiffen ye up ter move inter sech
quarters ez these." Dundas hesitated, but the
mountaineer had already taken assent for granted,
and pushed his horse into a sharp trot. Evidently
a refusal was not in order. Dundas pressed forward, and they rode together along the winding
way past the ten-pin alley, its long low roof half
hidden in the encroaching undergrowth springing
up apace beneath the great trees; past the stables;
past a line of summer cottages, strangely staring of
aspect out of the yawning doors and windows, giving, instead of an impression of vacancy, a sense
of covert watching, of secret occupancy. If one's
glances were only quick enough, were there not
faces pressed to those shattered panes — scarcely
seen—swiftly withdrawn?

He was in a desert; he had hardly been so
utterly alone in all his life; yet he bore through
the empty place a feeling of espionage, and ever
and anon he glanced keenly at the overgrown
lawns, with their deepening drifts of autumn leaves,
at the staring windows and flaring doors, which
emitted sometimes sudden creaking wails in the
silence, as if he sought to assure himself of the
vacancy of which his mind took cognizance and
yet all his senses denied.

Little of his sentiment, although sedulously
cloaked, was lost on Sim Roxby; and he was
aware, too, in some subtle way, of the relief his
guest experienced when they plunged into the
darkening forest and left the forlorn place behind
them. The clearing in which it was situated seemed

an oasis of light in the desert of night in which
the rest of the world lay. From the obscurity of
the forest Dundas saw, through the vistas of the
giant trees, the clustering cottages, the great hotel,
gables and chimneys and tower, stark and distinct as in some weird dream-light in the midst of
the encircling gloom. The after-glow of sunset
was still aflare on the western windows; the whole
empty place was alight with a reminiscence of its
old aspect—its old gay life. Who knows what
memories were a-stalk there—what semblance of
former times? What might not the darkness
foster, the impunity of desertion, the associations
that inhabited the place with almost the strength
of human occupancy itself? Who knows — who
knows?

He remembered the scene afterward, the impression he received. And from this, he thought,
arose his regret for his decision to take up here
his abiding-place.

The forest shut out the illumined landscape,
and the night seemed indeed at hand; the gigantic
boles of the trees loomed through the encompassing gloom, that was yet a semi-transparent medium, like some dark but clear fluid through which
objects were dimly visible, albeit tinged with its
own sombre hue. The lank, rawboned sorrel had
set a sharp pace, to which the chestnut, after momentary lagging, as if weary with the day's travel,
responded briskly. He had received in some way
intimations that his companion's corn-crib was
near at hand, and if he had not deduced from
these premises the probability of sharing his fare,

his mental processes served him quite as well as reason, and brought him to the same result. On and on they sped, neck and neck, through the darkening woods; fire flashed now and again from their iron-shod hoofs; often a splash and a shower of drops told of a swift dashing through the mud-holes that recent rains had fostered in the shallows. The dank odor of dripping boughs came on the clear air. Once the chestnut shied from a sudden strange shining point springing up in the darkness close at hand, which the country-bred horse discriminated as fox-fire, and kept steadily on, unmindful of the rotting log where it glowed. Far in advance, in the dank depths of the woods, a Will-o'-the-wisp danced and flickered and lured the traveller's eye. The stranger was not sure of the different quality of another light, appearing down a vista as the road turned, until the sorrel, making a tremendous spurt, headed for it, uttering a joyous neigh at the sight.

The deep-voiced barking of hounds rose melodiously on the silence, and as the horses burst out of the woods into a small clearing, Dundas beheld in the brighter light a half-dozen of the animals nimbly afoot in the road, one springing over the fence, another in the act of climbing, his fore-paws on the topmost rail, his long neck stretched, and his head turning about in attitudes of observation. He evidently wished to assure himself whether the excitement of his friends was warranted by the facts before he troubled himself to vault over the fence. Three or four still lingered near the door of a log-cabin, fawning about a girl who stood on the

porch. Her pose was alert, expectant; a fire in the
dooryard, where the domestic manufacture of soap
had been in progress, cast a red flare on the house,
its appurtenances, the great dark forest looming
all around, and, more than the glow of the hearth
within, lighted up the central figure of the scene.
She was tall, straight, and strong; a wealth of fair
hair was clustered in a knot at the back of her
head, and fleecy tendrils fell over her brow; on it
was perched a soldier's cap; and certainly more
gallant and fearless eyes had never looked out from
under the straight, stiff brim. Her chin, firm,
round, dimpled, was uplifted as she raised her
head, descrying the horsemen's approach. She
wore a full dark-red skirt, a dark brown waist, and
around her neck was twisted a gray cotton kerchief,
faded to a pale ashen hue, the neutrality of which
somehow aided the delicate brilliancy of the blended
roseate and pearly tints of her face. Was this the
seer of ghosts — Dundas marvelled — this the Millicent whose pallid and troubled phantom already
paced the foot-bridge? .

· He did not realize that he had drawn up his
horse suddenly at the sight of her, nor did he notice that his host had dismounted, until Roxby was
at the chestnut's head, ready to lead the animal to
supper in the barn. His evident surprise, his preoccupation, were not lost upon Roxby, however.
His hand hesitated on the girth of the chestnut's
saddle when he stood between the two horses in
the barn. He had half intended to disregard the
stranger's declination of his invitation, and stable
the creature. Then he shook his head slowly; the

mystery that hung about the new-comer was not reassuring. "A heap o' wuthless cattle 'mongst them valley men," he said; for the war had been in some sort an education to his simplicity. "Let him stay whar the cunnel expected him ter stay. I ain't wantin' no stranger a-hangin' round about Mill'cent, nohow. Em'ry Keenan ain't a pattern o' perfection, but I be toler'ble well acquainted with the cut o' his foolishness, an' I know his daddy an' mammy, an' both sets o' gran'daddies an' gran'mammies, an' I could tell ye exac'ly which one the critter got his nose an' his mouth from, an' them lean sheep's-eyes o' his'n, an' nigh every tone o' his voice. Em'ry never thunk afore ez I set store on bein' acquainted with him. He 'lowed I knowed him *too* well."

He laughed as he glanced through the open door into the darkening landscape. Horizontal gray clouds were slipping fast across the pearly spaces of the sky. The yellow stubble gleamed among the brown earth of the farther field, still striped with its furrows. The black forest encircled the little cleared space, and a wind was astir among the tree-tops. A white star gleamed through the broken clapboards of the roof, the fire still flared under the soap-kettle in the dooryard, and the silence was suddenly smitten by a high cracked old voice, which told him that his mother had perceived the dismounted stranger at the gate, and was graciously welcoming him.

She had come to the door, where the girl still stood, but half withdrawn in the shadow. Dundas silently bowed as he passed her, following his aged hostess into the low room, all bedight with the firelight of a huge chimney-place, and comfortable with the realization of a journey's end. The wilderness might stretch its weary miles around, the weird wind wander in the solitudes, the star look coldly on unmoved by aught it beheld, the moon show sad portents, but at the door they all failed, for here waited rest and peace and human companionship and the sense of home.

"Take a cheer, stranger, an' make yerself at home. Powerful glad ter see ye—war 'feard night would overtake ye. Ye fund the water toler'ble high in all the creeks an' sech, I reckon, an' fords shifty an' onsartin. Yes, sir. Fall rains kem on earlier'n common, an' more'n we need. Wisht we could divide it with that thar drought we had in the summer. Craps war cut toler'ble short, sir—toler'ble short."

Mrs. Roxby's spectacles beamed upon him with an expression of the utmost benignity as the firelight played on the lenses, but her eyes peering over them seemed endowed in some sort with independence of outlook. It was as if from behind some bland mask a critical observation was poised for unbiased judgment. He felt in some degree under surveillance. But when a light step heralded an approach he looked up, regardless of the betrayal of interest, and bent a steady gaze upon Millicent as she paused in the doorway.

And as she stood there, distinct in the firelight and outlined against the black background of the night, she seemed some modern half-military ideal of Diana, with her two gaunt hounds beside her,

the rest of the pack vaguely glimpsed at her heels
outside, the perfect outline and chiselling of her
features, her fine, strong, supple figure, the look
of steady courage in her eyes, and the soldier's
cap on her fair hair.   Her face so impressed itself
upon his mind that he seemed to have seen her
often.   It was some resemblance to a picture of a
vivandière, doubtless, in a foreign gallery—he could
not say when or where ; a remnant of a tourist's
overcrowded impressions ; a half-realized reminiscence, he thought, with an uneasy sense of recognition.

"Hello, Mill'cent ! home agin !" Roxby cried, in
cheery greeting as he entered at the back door opposite.   "What sorter topknot is that ye got on ?"
he demanded, looking jocosely at her head-gear.

The girl put up her hand with an expression of
horror.   A deep red flush dyed her cheek as she
touched the cap.   "I forgot 'twar thar," she murmured, contritely.   Then, with a sudden rush of
anger as she tore it off : "'Twar granny's fault.
She axed me ter put it on, so ez ter see which one
I looked most like."

"Stranger," quavered the old woman, with a
painful break in her voice, "I los' fower sons in
the war, an' Mill'cent hev got the fambly favor."

"Ye *mought* hev let me know ez I war a-perlitin'
round in this hyar men's gear yit," the girl muttered, as she hung the cap on a prong of the deer
antlers on which rested the rifle of the master of
the house.

Roxby's face had clouded at the mention of the
four sons who had gone out from the mountains

never to return, leaving to their mother's aching
heart only the vague comfort of an elusive resemblance in a girl's face; but as he noted Millicent's
pettish manner, and divined her mortification because of her unseemly head-gear in the stranger's
presence, he addressed her again in that jocose tone
without which he seldom spoke to her.

"Warn't you-uns apologizin' ter me t'other day
fur not bein' a nephew 'stiddier a niece?   Looked
sorter like a nephew ter-night."

She shook her head, covered now only with its
own charming tresses waving in thick undulations
to the coil at the nape of her neck — a trifle dishevelled from the rude haste with which the cap
had been torn off.

Roxby had seated himself, and with his elbows
on his knees he looked up at her with a teasing
jocularity, such as one might assume toward a
child.

"*Ye war*," he declared, with affected solemnity
—"ye war 'pologizin' fur not bein' a nephew, an'
'lowed ef ye war a nephew we could go a-huntin'
tergether, an' ye could holp me in all my quar'ls
an' fights.   I been aging some lately, an' ef I war
ter go ter the settlemint an' git inter a fight I
mought not be able ter hold my own.   Think what
'twould be ter a pore old man ter hev a dutiful
nephew step up an'"—he doubled his fists and
squared off — "jes' let daylight through some o'
them cusses.   An' didn't *ye say*"—he dropped his
belligerent attitude and pointed an insistent finger
at her, as if to fix the matter in her recollection—
"ef ye war a nephew 'stiddier a niece ye could fire

a gun 'thout shettin' yer eyes? An' I told ye then ez that would mend yer aim mightily. I told ye that I'd be powerful mortified ef I hed a nephew ez hed ter shet his eyes ter keep the noise out'n his ears whenst he fired a rifle. The tale would go mighty hard with me at the settlemint."

The girl's eyes glowed upon him with the fixity and the lustre of those of a child who is entertained and absorbed by an elder's jovial wiles. A flash of laughter broke over her face, and the low, gurgling, half-dreamy sound was pleasant to hear. She was evidently no more than a child to these bereft old people, and by them cherished as naught else on earth.

"An' didn't *I* tell *you-uns*," he went on, affecting to warm to the discussion, and in reality oblivious of the presence of the guest—"didn't I tell ye ez how ef ye war a nephew 'stiddier a niece ye wouldn't hev sech cattle ez Em'ry Keenan a-danglin' round underfoot, like a puppy ye can't gin away, an' that *won't* git lost, an' ye ain't got the heart ter kill?"

The girl's lip suddenly curled with scorn. "Yer nephew would be obligated ter make a ch'ice fur marryin' 'mongst these hyar mounting gals—Parmely Lepstone, or Belindy M'ria Matthews, or one o' the Windrow gals. Waal, sir, I'd ruther be yer niece—even ef Em'ry Keenan *air* like a puppy underfoot, that ye can't gin away, an' won't git lost, an' ye ain't got the heart ter kill." She laughed again, showing her white teeth. She evidently relished the description of the persistent adherence of poor Emory Keenan. "But which one o' these

hyar gals would ye recommend ter yer nephew ter
marry—ef ye hed a nephew?"

She looked at him with flashing eyes, conscious
of having propounded a poser.

He hesitated for a moment. Then—"I'm sur-
rounded," he said, with a laugh. "Ez I couldn't
find a wife fur myself, I can't ondertake ter recommend one ter my nephew. Mighty fine boy he'd
hev been, an' saaft-spoken an' perlite ter aged men
—not sassy an' makin' game o' old uncles like a
niece. Mighty fine boy!"

"Ye air welcome ter him," she said, with a sim-
ulation of scorn, as she turned away to the table.

Whether it were the military cap she had worn,
or the fancied resemblance to the young soldiers,
never to grow old, who had gone forth from this
humble abode to return no more, there was still to
the guest's mind the suggestion of the vivandière
about her as she set the table and spread upon it
the simple fare. To and from the fireplace she was
followed by two or three of the younger dogs, their
callowness expressed in their lack of manners and
perfervid interest in the approaching meal. This
induced their brief journeys back and forth, albeit
embarrassed by their physical conformation, short
turns on four legs not being apparently the easy
thing it would seem from so much youthful supple-
ness. The dignity of the elder hounds did not
suffer them to move, but they looked on from erect
postures about the hearth with glistening eyes and
slobbering jaws.

Ever and anon the deep blue eyes of Millicent
were lifted to the outer gloom, as if she took note

of its sinister aspect. She showed scant interest in
the stranger, whose gaze seldom left her as he sat
beside the fire. He was a handsome man, his
face and figure illumined by the firelight, and it
might have been that he felt a certain pique, an
unaccustomed slight, in that his presence was so
indifferent an element in the estimation of any
young and comely specimen of the feminine sex.
Certainly he had rarely encountered such absolute
preoccupation as her smiling far-away look betokened as she went back and forth with her young
canine friends at her heels, or stood at the table
deftly slicing the salt-rising bread, the dogs poised
skilfully upon their hind-legs to better view the ap-
petizing performance ; whenever she turned her
face toward them they laid their heads languish-
ingly askew, as if to remind her that supper could
not be more fitly bestowed than on them. One, to
steady himself, placed unobserved his fore-paw on
the edge of the table, his well-padded toes leaving
a vague imprint as of fingers upon the coarse white
cloth; but John Dundas was a sportsman, and
could the better relax an exacting nicety where
so pleasant-featured and affable a beggar was concerned. He forgot the turmoils of his own troubles
as he gazed at Millicent, the dreary aspect of the
solitudes without, the exile from his accustomed
sphere of culture and comfort, the poverty and
coarseness of her surroundings. He was sorry
that he had declined a longer lease of Roxby's
hospitality, and it was in his mind to reconsider
when it should be again proffered. Her attitude,
her gesture, her face, her environment, all appealed

to his sense of beauty, his interest, his curiosity,
as little ever had done heretofore.   Slice after slice
of the firm fragrant bread was deftly cut and laid
on the plate, as again and again she lifted her eyes
with a look that might seem to expect to rest on
summer in the full flush of a June noontide without, rather than on the wan, wintry night sky and
the plundered, quaking woods, while the robber
wind sped on his raids hither and thither so swiftly
that none might follow, so stealthily that none might
hinder.   A sudden radiance broke upon her face, a
sudden shadow fell on the firelit floor, and there
was entering at the doorway a tall, lithe young
mountaineer, whose first glance, animated with a
responsive brightness, was for the girl, but whose
punctilious greeting was addressed to the old
woman.

"Howdy, Mis' Roxby—howdy?   Air yer rheumatics mendin' enny?" he demanded, with the con-
dolent suavity of the would-be son-in-law, or grand-
son-in-law, as the case may be.   And he hung with
a transfixed interest upon her reply, prolix and discursive according to the wont of those who cultivate
"rheumatics," as if each separate twinge racked
his own sympathetic and filial sensibilities.   Not
until the tale was ended did he set his gun against
the wall and advance to the seat which Roxby had
indicated with the end of the stick he was whittling.
He observed the stranger with only slight interest,
till Dundas drew up his chair opposite at the table.
There the light from the tallow dip, guttering in
the centre, fell upon his handsome face and eyes,
his carefully tended beard and hair, his immaculate

cuffs and delicate hand, the seal-ring on his taper
finger.

"Like a gal, by gum!" thought Emory Keenan.
"Rings on his fingers—yit six feet high!"

He looked at his elders, marvelling that they so
hospitably repressed the disgust which this effeminate adornment must occasion, forgetting that it
was possible that they did not even observe it.   In
the gala-days of the old hotel, before the war, they
had seen much "finicking finery" in garb and equi-
page and habits affected by the *jeunesse dorée* who
frequented the place in those halcyon times, and
were accustomed to such details. . It might be that
they and Millicent approved such flimsy daintiness.
He began to fume inwardly with a sense of infe-
riority in her estimation.   One of his fingers had
been frosted last winter, and with the first twinge
of cold weather it was beginning to look very red
and sad and clumsy, as if it had just remembered
its ancient woe ; he glanced from it once more at
the delicate ringed hand of the stranger.

Dundas was looking up with a slow, deferential,
decorous smile that nevertheless lightened and
transfigured his expression.   It seemed somehow
communicated to Millicent's face as she looked
down at him from beneath her white eyelids and
long, thick, dark lashes, for she was standing beside him, handing him the plate of bread.   Then,
still smiling, she passed noiselessly on to the others.

Emory was indeed clumsy, for he had stretched
his hand downward to offer a morsel to a friend of
his under the table—he was on terms of exceeding
amity with the four-footed members of the house

3

hold—and in his absorption not withdrawing it as swiftly as one accustomed to canine manners should do, he had his frosted finger well mumbled before he could, as it were, repossess himself of it.

"I wonder what they charge fur iron over yander at the settlemint, Em'ry?" observed Sim Roxby presently.

"Dun'no', sir," responded Emory, glumly, his sullen black eyes full of smouldering fire—" hevin' no call ter know, ez I ain't no blacksmith."

"I war jes' wonderin' ef tenpenny nails didn't cost toler'ble high ez reg'lar feed," observed Roxby, gravely.

But his mother laughed out with a gleeful cracked treble, always a ready sequence of her son's rustic sallies. "He got ye that time, Em'ry," she cried.

A forced smile crossed Emory's face. He tossed back his tangled dark hair with a gasp that was like the snort of an unruly horse submitting to the inevitable, but with restive projects in his brain. "I let the dog hyar ketch my finger whilst feedin' him," he said. His plausible excuse for the tenpenny expression was complete; but he added, his darker mood recurring instantly, "An', Mis' Roxby, I hev put a stop ter them ez hev tuk ter callin' me Em'ly, I hev."

The old woman looked up, her small wrinkled mouth round and amazed. "*I* never called ye Emily," she declared.

Swift repentance seized him.

"Naw, 'm," he said, with hurried propitiation. "I 'lowed ye did."

"I didn't," said the old woman. "But ef I war

ter find it toothsome ter call ye 'Emily,' I dun'no'
how ye air goin' ter pervent it.   Ye can't go gun-
nin' fur me, like ye done fur the men at the mill, fur
callin' ye 'Emily.'"

"Law, Mis' Roxby!" he could only exclaim, in
his horror and contrition at this picture he had
thus conjured up.   "Ye air welcome ter call me
ennything ye air a mind ter," he protested.

And then he gasped once more.   The eyes of
the guest, contemptuous, amused, seeing through
him, were fixed upon him.   And he himself had
furnished the lily-handed stranger with the information that he had been stigmatized "Em'ly" in the
banter of his associates, until he had taken up arms,
as it were, to repress this derision.

"It takes powerful little ter put ye down, Em'ry,"
said Roxby, with rallying laughter.   "Mam hev sent
ye skedaddlin' in no time at all.   I don't b'lieve the
Lord made woman out'n the man's rib.   He made
her out'n the man's backbone; fur the man ain't
hed none ter speak of sence."

Millicent, with a low gurgle of laughter, sat down
beside Emory at the table, and fixed her eyes, softly lighted with mirth, upon him.   The others too
had laughed, the stranger with a flattering intonation, but young Keenan looked at her with a dumb
appealing humility that did not altogether fail of
its effect, for she busied herself to help his plate
with an air of proprietorship as if he were a child,
and returned it with a smile very radiant and sufficient at close range.   She then addressed herself
to her own meal.   The young dogs under the table ceased to beg, and gambolled and gnawed and

tugged at her stout little shoes, the sound of their callow mirthful growls rising occasionally above the talk. Sometimes she rose again to wait on the table, when they came leaping out after her, jumping and catching at her skirts, now and then casting themselves on the ground prone before her feet, and rolling over and over in the sheer joy of existence.

The stranger took little part in the talk at the table. Never a question was asked him as to his mission in the mountains, or the length of his stay, his vocation, or his home. That extreme courtesy of the mountaineers, exemplified in their singular abstinence from any expressions of curiosity, accepted such account of himself as he had volunteered, and asked for no more. In the face of this standard of manners any inquisitiveness on his part, such as might have elicited points of interest for his merely momentary entertainment, was tabooed. Nevertheless, silent though he was for the most part, the relish with which he listened, his half-covert interest in the girl, his quick observation of the others, the sudden very apparent enlivening of his mental atmosphere, betokened that his quarters were not displeasing to him. It seemed only a short time before the meal was ended and the circle all, save Millicent, with pipes alight before the fire again. The dogs, well fed, had ranged themselves on the glowing hearth, lying prone on the hot stones ; one old hound, however, who conserved the air of listening to the conversation, sat upright and nodded from time to time, now and again losing his balance and tipping forward in a truly human fashion, then

gazing round on the circle with an open luminous
eye, as who should say he had not slept.

It was all very cheerful within, but outside the
wind still blared mournfully. Once more Dundas
was sorry that he had declined the invitation to remain, and it was with a somewhat tentative intention that he made a motion to return to the hotel.
But his host seemed to regard his resolution as
final, and rose with a regret, not an insistence. The
two women stared in silent amazement at the mere
idea of his camping out, as it were, in the old hotel.
The ascendency of masculine government here, not-
withstanding Roxby's assertion that Eve was made
of Adam's backbone, was very apparent in their
mute acquiescence and the alacrity with which they
began to collect various articles, according to his di-
rections, to make the stranger's stay more comfortable.

"Em'ry kin go along an' holp," he said, heart-
lessly; for poor Emory's joy in perceiving that the
guest was not a fixture, and that his presence was
not to be an embargo on any word between himself
and Millicent during the entire evening, was pitia-
bly manifest. But the situation was still not without
its comforts, since Dundas was to go too. Hence
he was not poor company when once in the saddle,
and was civil to a degree of which his former dismayed surliness had given no promise.

Night had become a definite element. The twilight had fled. Above their heads, as they galloped
through the dank woods, the bare boughs of the
trees clashed together—so high above their heads
that to the town man, unaccustomed to these great

growths, the sound seemed not of the vicinage, but unfamiliar, uncanny, and more than once he checked his horse to listen. As they approached the mountain's verge and overlooked the valley and beheld the sky, the sense of the predominance of darkness was redoubled. The ranges gloomed against the clearer spaces, but a cloud, deep gray with curling white edges, was coming up from the west, with an invisible convoy of vague films, beneath which the stars, glimmering white points, disappeared one by one. The swift motion of this aerial fleet sailing with the wind might be inferred from the seemingly hurried pace of the moon making hard for the west. Still bright was the illumined segment, but despite its glitter the shadowy space of the full disk was distinctly visible, its dusky field spangled with myriads of minute, dully golden points. Down, down it took its way in haste—in disordered fright, it seemed, as if it had no heart to witness the storm which the wind and the clouds foreboded—to fairer skies somewhere behind those western mountains. Soon even its vague light would encroach no more upon the darkness. The great hotel would be invisible, annihilated as it were in the gloom, and not even thus dimly exist, glimmering, alone, forlorn, so incongruous to the wilderness that it seemed even now some mere figment of the brain, as the two horsemen came with a freshened burst of speed along the deserted avenue and reined up beside a small gate at the side.

"No use ter ride all the way around," observed Emory Keenan. "Mought jes ez well 'light an' hitch hyar."

The moon gave him the escort of a great grotesque shadow as he threw himself from his horse
and passed the reins over a decrepit hitching-post
near at hand. Then he essayed the latch of the
small gate. He glanced up at Dundas, the moonlight in his dark eyes, with a smile as it resisted
his strength.

He was a fairly good-looking fellow when rid of
the self-consciousness of jealousy. His eyes, mouth,
chin, and nose, acquired from reliable and recognizable sources, were good features, and statuesque
in their immobility beneath the drooping curves of
his broad soft hat. He was tall, with the slen-
derness of youth, despite his evident weight and
strength. He was long-waisted and lithe and small
of girth, with broad square shoulders, whose play
of muscles as he strove with the gate was not altogether concealed by the butternut jeans coat belted
in with his pistols by a broad leathern belt. His
boots reached high on his long legs, and jingled
with a pair of huge cavalry spurs. His stalwart
strength seemed as if it must break the obdurate
gate rather than open it, but finally, with a rasping
creak, dismally loud in the silence, it swung slowly
back.

The young mountaineer stood gazing for a moment at the red rust on the hinges. "How long
sence this gate must hev been opened afore?" he
said, again looking up at Dundas with a smile.

Somehow the words struck a chill to the stranger's heart. The sense of the loneliness of the
place, of isolation, filled him with a sort of awe.
The night-bound wilderness itself was not more

daunting than these solitary tiers of piazzas, these vacant series of rooms and corridors, all instinct with vanished human presence, all alert with echoes of human voices. A step, a laugh, a rustle of garments—he could have sworn he heard them at any open doorway as he followed his guide along the dim moonlit piazza, with its pillars duplicated at regular intervals by the shadows on the floor. How their tread echoed down these lonely ways! From the opposite side of the house he heard Keenan's spurs jangling, his soldierly stride sounding back as if their entrance had roused barracks. He winced once to see his own shadow with its stealthier movement. It seemed painfully furtive. For the first time during the evening his jaded mind, that had instinctively sought the solace of contemplating trifles, reverted to its own tormented processes. "Am I not hiding?" he said to himself, in a sort of sarcastic pity of his plight.

The idea seemed never to enter the mind of the transparent Keenan. He laughed out gayly as they turned into the weed-grown quadrangle, and the red fox that Dundas had earlier observed slipped past him with affrighted speed and dashed among the shadows of the dense shrubbery of the old lawn without. Again and again the sound rang back from wall to wall, first with the jollity of seeming imitation, then with an appalled effect sinking to silence, and suddenly rising again in a grewsome *staccato* that suggested some terrible unearthly laughter, and bore but scant resemblance to the hearty mirth which had evoked it. Keenan paused and looked back with friendly gleaming eyes. "Oughter been a leetle handier with these hyar consarns," he said, touching the pistols in his belt.

It vaguely occurred to Dundas that the young man went strangely heavily armed for an evening visit at a neighbor's house. But it was a lawless country and lawless times, and the sub-current of suggestion .did not definitely fix itself in his mind until he remembered it later. He was looking into each vacant open doorway, seeing the still moonlight starkly white upon the floor; the cobwebbed and broken window-panes, through which a section of leafless trees beyond was visible; bits of furniture here and there, broken by the vandalism of the guerillas. Now and then a scurrying movement told of a gopher, hiding too, and on one mantel-piece, the black fireplace yawning below, sat a tiny tawny-tinted owl, whose motionless bead-like eyes met his with a stare of stolid surprise. After he had passed, its sudden ill-omened cry set the silence to shuddering.

Keenan, leading the way, paused in displeasure. "I wisht I hed viewed that critter," he said, glumly. "I'd hev purvented that screechin' ter call the devil, sure. It's jes a certain sign o' death."

He was about to turn, to wreak his vengeance, perchance. But the bird, sufficiently fortunate itself, whatever woe it presaged for others, suddenly took its awkward flight through sheen and shadow across the quadrangle, and when they heard its cry again it came from some remote section of the building, with a doleful echo as a refrain.

The circumstance was soon forgotten by Kee-

nan.   He seemed a happy, mercurial, lucid nature,
and he began presently to dwell with interest on
the availability of the old music-stand in the centre
of the square as a manger.   "Hyar," he said, striking the rotten old structure with a heavy hand, which
sent a quiver and a thrill through all the timbers—
"hyar's whar the guerillas always hitched thar
beastises.   Thar feed an' forage war piled up thar
on the fiddlers' seats.   Ye can't do no better'n ter
pattern arter them, till ye git ready ter hev fiddlers
an' sech a-sawin' away in hyar agin."

And he sauntered away from the little pavilion,
followed by Dundas, who had not accepted his suggestion of a room on the first floor as being less
liable to leakage, but finally made choice of an inner
apartment in the second story.   He looked hard
at Keenan, when he stood in the doorway surveying the selection.   The room opened into a cross-
hall which gave upon a broad piazza that was lat-
ticed; tiny squares of moonlight were all sharply
drawn on the floor, and, seen through a vista of
gray shadow, seemed truly of a gilded lustre.
From the windows of this room on a court-yard
no light could be visible to any passer-by without.
Another door gave on an inner gallery, and through
its floor a staircase came up from the quadrangle
close to the threshold.   Dundas wondered if these
features were of possible significance in Keenan's
estimation.   The young mountaineer turned suddenly, and snatching up a handful of slats broken
from the shutters, remarked:

"Let's see how the chimbly draws—that's the
main p'int."

There was no defect in the chimney's constitution. It drew admirably, and with the white and red flames dancing in the fireplace, two or three chairs, more or less disabled, a table, and an uphol-stered lounge gathered at random from the rooms near at hand, the possibility of sojourning comfort-ably for a few days in the deserted hostelry seemed amply assured.

Once more Dundas gazed fixedly at the face of the young mountaineer, who still bent on one knee on the hearth, watching with smiling eyes the tri-umphs of his fire-making. It seemed to him after-wards that his judgment was strangely at fault: he perceived naught of import in the shallow bright-ness of the young man's eyes, like the polished sur-face of jet; in the instability of his jealousy, his anger; in his hap-hazard, mercurial temperament. Once he might have noted how flat were the spaces beneath the eyes, how few were the lines that de-fined the lid, the socket, the curve of the cheek-bone, the bridge of the nose, and how expression-less. It was doubtless the warmth and glow of the fire, the clinging desire of companionship, the earnest determination to be content, pathetic in one who had but little reason for optimism, that caused him to ignore the vacillating glancing moods that successively swayed Keenan, strong while they lasted, but with scanty augury because of their evanescence. He was like some newly discovered property in physics of untried poten-tialities, of which nothing is ascertained but its un-certainties.

And yet he seemed to Dundas a simple country

fellow, good-natured in the main, unsuspicious, and helpful. So, giving a long sigh of relief and fatigue, Dundas sank down in one of the large arm-chairs that had once done duty for the summer loungers on the piazza.

In the light of the fire Emory was once more looking at him. A certain air of distinction, a grace and ease of movement, an indescribable quality of bearing which he could not discrimi-nate, yet which he instinctively recognized as su-perior, offended him in some sort. He noticed again the ring on the stranger's hand as he drew off his glove. Gloves! Emory Keenan would as soon have thought of wearing a petticoat. Once more the fear that these effeminate graces found favor in Millicent's estimation smote upon his heart. It made the surface of his opaque eyes glisten as Dundas rose and took up a pipe and to-bacco-pouch which he had laid on the mantel-piece, his full height and fine figure shown in the changed posture.

"Ez tall ez me, ef not taller, an', by gum! a good thirty pound heavier," Emory reflected, with a growing dismay that he had not those stalwart claims to precedence in height and weight as an offset to the smoother fascinations of the stranger's polish.

He had risen hastily to his feet. He would not linger to smoke fraternally over the fire, and thus cement friendly relations.

"I guided him hyar, like old Sim Roxby axed me ter do, an' that's all. I ain't keerin' ef I never lay eyes on him again," he said to himself.

"Going?" said Dundas, pleasantly, noticing the motion. "You'll look in again, won't you?"

"Wunst in a while, I reckon," drawled Keenan, a trifle thrown off his balance by this courtesy.

He paused at the door, looking back over his shoulder for a moment at the illumined room, then stepped out into the night, leaving the tenant of the lonely old house filling his pipe by the fire.

His tread rang along the deserted gallery, and sudden echoes came tramping down the vacant halls as if many a denizen of the once populous place was once more astir within its walls. Long after Dundas had heard him spring from the lower piazza to the ground, and the rusty gate clang behind him, vague footfalls were audible far away, and were still again, and once more a pattering tread in some gaunt and empty apartment near at hand, faint and fainter yet, till he hardly knew whether it were the reverberations of sound or fancy that held his senses in thrall.

And when all was still and silent at last he felt less solitary than when these elusive tokens of human presence were astir.

Late, late he sat over the dwindling embers. His mind, no longer diverted by the events of the day, recurred with melancholy persistence to a theme which even they, although fraught with novelty and presage of danger, had not altogether crowded out. And as the sense of peril dulled, the craft of sophistry grew clumsy. Remorse laid hold upon him in these dim watches of the night. Self-reproach had found him out here, defenceless so far from the specious wiles and ways of men. All the line of provo-

cations seemed slight, seemed naught, as he reviewed
them and balanced them against a human life. True,
it was not in some mad quarrel that his skill had
taken it and had served to keep his own—a duel, a
fair fight, strictly regular according to the code of
" honorable men " for ages past—and he sought to
argue that it was doubtless but the morbid sense of
the wild fastnesses without, the illimitable vastness
of the black night, the unutterable indurability of nature to the influences of civilization, which made it
taste like murder. He had brought away even from
the scene of action, to which he had gone with decorous deliberation — his worldly affairs arranged
for the possibility of death, his will made, his volition surrendered, and his sacred honor in the hands
of his seconds — a humiliating recollection of the
sudden revulsion of the aspect of all things ; the
criminal sense of haste with which he was hurried
away after that first straight shot ; the agitation,
nay, the fright of his seconds ; their eagerness to be
swiftly rid of him, their insistence that he should go
away for a time, get out of the country, out of the
embarrassing purview of the law, which was prone
to regard the matter as he himself saw it now,
and which had an ugly trick of calling things by
their right names in the sincere phraseology of
an indictment. And thus it was that he was here,
remote from all the usual lines of flight, with his
affectation of being a possible purchaser for the
old hotel, far from the railroad, the telegraph,
even the postal service. Some time — soon, indeed, it might be, when the first flush of excitement and indignation should be overpast, and the

law, like a barking dog that will not bite, should
have noisily exhausted the gamut of its devoirs—he
would go back and live according to his habit in his
wonted place, as did other men whom he had known
to be "called out," and who had survived their op-
ponents.   Meantime he heard the ash crumble ; he
saw the lighted room wane from glancing yellow to
a dull steady red, and so to dusky brown ; he marked
the wind rise, and die away, and come again, bang-
ing the doors of the empty rooms, and setting timbers all strangely to creaking as under sudden trampling feet; then lift into the air with a rustling sound
like the stir of garments and the flutter of wings,
calling out weirdly in the great voids of the upper
atmosphere.

He had welcomed the sense of fatigue earlier in
the evening, for it promised sleep.   Now it had
slipped away from him.   He was strong and young,
and the burning sensation that the frosty air had
left on his face was the only token of the long journey.   It seemed as if he would never sleep again
as he lay on the lounge watching the gray ash gradually overgrow the embers, till presently only a
vague dull glow gave intimation of the position of
the hearth in the room.   And then, bereft of this
dim sense of companionship, he stared wide-eyed
in the darkness, feeling the only creature alive and
awake in all the world.   No ; the fox was suddenly
barking within the quadrangle — a strangely wild
and alien tone.   And presently he heard the animal trot past his door on the piazza, the cushioned
footfalls like those of a swift dog.   He thought with
a certain anxiety of the tawny tiny owl that had sat

like a stuffed ornament on the mantel-piece of a neighboring room, and he listened with a quaking vicarious presentiment of woe for the sounds of capture and despair.   He was sensible of waiting and hoping for the fox's bootless return, when he suddenly lost consciousness.

How long he slept he did not know, but it seemed only a momentary respite from the torture of memory, when, still in the darkness, thousands of tremulous penetrating sounds were astir, and with a great start he recognized the rain on the roof. It was coming down in steady torrents that made the house rock before the tumult of his plunging heart was still, and he was longing again for the forgetfulness of sleep.   In vain.   The hours dragged by; the windows slowly, slowly defined their dull gray squares against the dull gray day dawning without.   The walls that had been left with only . the first dark coat of plaster, awaiting another season for the final decoration, showed their drapings of cobweb, and the names and pencilled scribblings with which the fancy of transient bushwhackers had chosen to deface them.   The locust-trees within the quadrangle drearily tossed their branches to and fro in the wind, the bark very black and distinct against the persistent gray lines of rain and the white walls of the galleried buildings opposite ; the gutters were brimming, roaring along like miniature torrents; nowhere was the fox or the owl to be seen.   Somehow their presence would have been a relief—the sight of any living thing reassuring. As he walked slowly along the deserted piazzas, in turning sudden corners, again and again he

paused, expecting that something, some one, was
approaching to meet him. When at last he mounted his horse, that had neighed gleefully to see
him, and rode away through the avenue and along
the empty ways among the untenanted summer cottages, all the drearier and more forlorn because of
the rain, he felt as if he had left an aberration,
some hideous dream, behind, instead of the stark
reality of the gaunt and vacant and dilapidated
old house.

The transition to the glow and cheer of Sim
Roxby's fireside was like a rescue, a restoration.
The smiling welcome in the women's eyes, their
soft drawling voices, with mellifluous intonations
that gave a value to each commonplace simple
word, braced his nerves like a tonic. It might have
been only the contrast with the recollections of the
night, with the prospect visible through the open
door—the serried lines of rain dropping aslant from
the gray sky and elusively outlined against the dark
masses of leafless woods that encircled the clearing; the dooryard half submerged with puddles of
a clay-brown tint, embossed always with myriads of
protruding drops of rain, for however they melted
away the downpour renewed them, and to the eye
they were stationary, albeit pervaded with a continual tremor—but somehow he was cognizant of a
certain coddling tenderness in the old woman's
manner that might have been relished by a petted
child, an unaffected friendliness in the girl's clear
eyes. They made him sit close to the great wood
fire; the blue and yellow flames gushed out from
the piles of hickory logs, and the bed of coals

4

gleamed at red and white heat beneath. They took
his hat to carefully dry it, and they spread out his
cloak on two chairs at one side of the room, where
it dismally dripped. When he ventured to sneeze,
Mrs. Roxby compounded and administered a "yerb
tea," a sovereign remedy against colds, which he
tasted on compulsion and in great doubt, and
swallowed with alacrity and confidence, finding its
basis the easily recognizable "toddy." He had
little knowledge how white and troubled his face
had looked às he came in from the gray day, how
strongly marked were those lines of sharp mental
distress, how piteously apparent was his mute appeal for sympathy and comfort.

"Mill'cent," said the old woman in the shed-room,
as they washed and wiped the dishes after the cozy
breakfast of venison and corn-dodgers and honey
and milk, "that thar man hev run agin the law,
sure's ye air born."

Millicent turned her reflective fair face, that
seemed whiter and more delicate in the damp
dark day, and looked doubtfully out over the
fields, where the water ran in steely lines in the
furrows.

"Mus' hev been by accident or suthin'. *He* ain't
no hardened sinner."

"Shucks!" the old woman commented upon her
reluctant acquiescence. "I ain't keerin' for the
law! 'Tain't none o' my job. The tomfool men
make an' break it. Ennybody ez hev seen this war
air obleeged to take note o' the wickedness o' men
in gineral. This hyer man air a sorter pitiful sinner, an' he hev got a look in his eyes that plumb

teches my heart. I 'ain't got no call ter know
nuthin' 'bout the law, bein' a 'oman an' naterally
ignorunt. I dun'no' ez he hev run agin it."

"Mus' hev been by accident," said Millicent,
dreamily, still gazing over the sodden fields.

The suspicion did nothing to diminish his comfort or their cordiality. The morning dragged by
without change in the outer aspects. The noontide
dinner came and went without Roxby's return, for the
report of the washing away of a bridge some miles
distant down the river had early called him out to the
scene of the disaster, to verify in his own interests
the rumor, since he had expected to haul his wheat
to the settlement the ensuing day. The afternoon
found the desultory talk still in progress about the
fire, the old woman alternately carding cotton and
nodding in her chair in the corner; the dogs eying the stranger, listening much of the time with
the air of children taking instruction, only occasionally wandering out-of-doors, the floor here
and there bearing the damp imprint of their feet;
and Millicent on her knees in the other corner, the
firelight on her bright hair, her delicate cheek, her
quickly glancing eyes, as she deftly moulded bullets.

"Uncle Sim hed ter s'render his shootin'-irons,"
she explained, "an' he 'ain't got no ca'tridge-loadin'
ones lef'. So he makes out with his old muzzle-
loadin' rifle that he hed afore the war, an' I moulds
his bullets for him rainy days."

As she held up a moulded ball and dexterously
clipped off the surplus lead, the gesture was so
culinary in its delicacy that one of the dogs in

front of the fire extended his head, making a long
neck, with a tentative sniff and a glistening glut-
tonous eye.

" Ef I swallered enny mo' lead, I wouldn't take
it hot, Towse," she said, holding out the bullet for
canine inspection.   " 'Tain't healthy !"

But the dog, perceiving the nature of the commodity, drew back with a look of deep reproach,
rose precipitately, and with a drooping tail went out
skulkingly into the wet gray day.

" Towse can't abide a bullet," she observed, "nor
nuthin' 'bout a gun.   He got shot wunst a-huntin',
an' he never furgot it.   Jes show him a gun an' he
ain't nowhar ter be seen—like he war cotch up in
the clouds."

" Good watch-dog, I suppose," suggested Dundas, striving to enter into the spirit of her talk.

" Naw ; too sp'ilt for a gyard-dog—granny cod-
dled him so whenst he got shot.   He's jest vally'ble
fur his conversation, I reckon," she continued, with
a smile in her eyes.   " I dun'no' what else, but he
is toler'ble good company."

The other dogs pressed about her, the heads of
the great hounds as high as her own as she sat
among them on the floor.   With bright eyes and
knitted brows they followed the motions of pouring
in the melted metal, the lifting of the bullets from
the mould, the clipping off of the surplus lead, and
the flash of the keen knife.

Outside the sad light waned ; the wind sighed
and sighed ; the dreary rain fell ; the trees clashed
their boughs dolorously together, and their turbulence deadened the sound of galloping horses.   As

Dundas sat and gazed at the girl's intent head, with its fleecy tendrils and its massive coil, the great hounds beside her, all emblazoned by the firelight upon the brown wall near by, with the vast fireplace at hand, the whole less like reality than some artist's pictured fancy, he knew naught of a sudden entrance, until she moved, breaking the spell, and looked up to meet the displeasure in Roxby's eyes and the dark scowl on Emory Keenan's face.

That night the wind shifted to the north. Morning found the chilled world still, ice where the water had lodged, all the trees incased in glittering garb that followed the symmetry alike of every bough and the tiniest twig, and made splendid the splintered remnants of the lightning-riven. The fields were laced across from furrow to furrow, in which the frozen water still stood gleaming, with white arabesques which had known a more humble identity as stubble and crab-grass; the sky was slate-colored, and from its sad tint this white splendor gained added values of contrast. When the sun should shine abroad much of the effect would be lost in the too dazzling glister; but the sun did not shine.

All day the gray mood held unchanged. Night was imperceptibly sifting down upon all this whiteness, that seemed as if it would not be obscured, as if it held within itself some property of luminosity, when Millicent, a white apron tied over her golden head, improvising a hood, its superfluous fulness gathered in many folds and pleats around her neck, fichu-wise, stood beside the ice-draped fodder-stack and essayed with half-numbed hands to insert a tallow dip into the socket of a lantern, all incrusted and clumsy with previous drippings.

"I dun'no' whether I be a-goin' ter need this hyar consarn whilst milkin' or no," she observed, half to herself, half to Emory, who, chewing a straw, somewhat surlily had followed her out for a word apart. "The dusk 'pears slow ter-night, but Spot's mighty late comin' home, an' old Sue air fractious an' contrairy-minded, and feels mighty anxious an' oneasy 'boutn her calf, that's ez tall ez she is nowa-days, an' don't keer no mo' 'bout her mammy 'n a half-grown human does. I tell her she oughtn't ter be mad with me, but with the way she brung up her chile, ez won't notice her now."

She looked up with a laugh, her eyes and teeth gleaming; her golden hair still showed its color be-neath the spotless whiteness of her voluminous head-gear, and the clear tints of her complexion seemed all the more delicate and fresh in the snowy pallor of the surroundings and the grayness of the evening.

"I reckon I'd better take it along," and once more she addressed herself to the effort to insert the dip into the lantern.

Emory hardly heard. His pulse was quick. His eye glittered. He breathed hard as, with both hands in his pockets, he came close to her.

"Mill'cent," he said, "I told ye the t'other day ez ye thunk a heap too much o' that thar stran-ger—"

"An' I tole ye, bubby, that I didn't think nuthin' o' nobody but you-uns," she interrupted, with an effort to placate his jealousy. The little jocularity which she affected dwindled and died before the

steady glow of his gaze, and she falteringly looked
at him, her unguided hands futilely fumbling with
the lantern.

"Ye can't fool me," he stoutly asseverated.
"Ye think mo' o' him 'n o' me, kase ye 'low he air
rich, an' book-larned, an' smooth-fingered, an' fini-
fied ez a gal, an' goin' ter buy the hotel.   I say,
hotel! Now *I'll* tell ye what he is — I'll tell ye !
He's a criminal. He's runnin' from the law. He's
hidin' in the old hotel that he's purtendin' ter buy."

She stared wide-eyed and pallid, breathless and
waiting.

He interpreted her expression as doubt, denial.

"It's gospel sure," he cried.   "Fur this very
evenin' I met a gang o' men an' the sheriff's deputy
down yander by the sulphur spring 'bout sundown,
an' he 'lowed ez they war a-sarchin' fur a criminal
ez war skulkin' round hyarabout lately — ez they
wanted a man fur hevin' c'mitted murder."

"But ye didn't accuse *him*, surely; ye hed no
right ter s'picion *him*. Uncle Sim ! Oh, my Lord !
Ye surely wouldn't ! Oh, Uncle Sim !"

Her tremulous words broke into a quavering cry
as she caught his arm convulsively, for his face con-
firmed her fears. She thrust him wildly away, and
started toward the house.

"Ye needn't go tattlin' on me," he said, roughly
pushing her aside. "I'll tell Mr. Roxby myself. I
ain't 'shamed o' what I done. I'll tell him. I'll
tell him myself." And animated with this intention
to forestall her disclosure, his long strides bore him
swiftly past and into the house.

It seemed to him that he lingered there only a

moment or two, for Roxby was not at the cabin, and
he said nothing of the quarrel to the old woman.
Already his heart had revolted against his treachery, and then there came to him the further reflection that he did not know enough to justify suspicion. Was not the stranger furnished with the
fullest credentials — a letter to Roxby from the
Colonel? Perhaps he had allowed his jealousy to
endanger the man, to place him in jeopardy even of
his life should he resist arrest.

Keenan tarried at the house merely long enough
to devise a plausible excuse for his sudden excited
entrance, and then took his way back to the barn-
yard.

It was vacant. The cows still stood lowing at
the bars ; the sheep cowered together in their shed ;
the great whitened cone of the fodder-stack gleamed
icily in the purple air; beside it lay the lantern
where Millicent had cast it aside. She was gone !
He would not believe it till he had run to the
barn, calling her name in the shadowy place, while
the horse at his manger left his corn to look over
the walls of his stall with inquisitive surprised eyes,
luminous in the dusk. He searched the hen-house,
where the fowls on their perches crowded close because of the chill of the evening. He even ran to
the bars and looked down across the narrow ravine
to which the clearing sloped. Beyond the chasm-
like gorge he saw presently on the high ascent opposite footprints that had broken the light frost-
like coating of ice on the dead leaves and moss —
climbing footprints, swift, disordered. He looked
back again at the lantern where Millicent had flung

it in her haste. Her mission was plain now. She had gone to warn Dundas. She had taken a direct line through the woods. She hoped to forestall the deputy sheriff and his posse, following the circuitous mountain road.

Keenan's lip curled in triumph. His heart burned hot with scornful anger and contempt of the futility of her effort. "They're there afore she started!" he said, looking up at the aspects of the hour shown by the sky, and judging of the interval since the encounter by the spring. Through a rift in the gray cloud a star looked down with an icy scintillation and disappeared again. He heard a branch in the woods snap beneath the weight of ice. A light sprang into the window of the cabin hard by, and came in a great gush of orange-tinted glow out into the snowy bleak wintry space. He suddenly leaped over the fence and ran like a deer through the woods.

Millicent too had been swift. He had thought to overtake her before he emerged from the woods into the more open space where the hotel stood. In this quarter the cloud-break had been greater. Toward the west a fading amber glow still lingered in long horizontal bars upon the opaque gray sky. The white mountains opposite were hung with purple shadows borrowed from a glimpse of sunset somewhere far away over the valley of East Tennessee; one distant lofty range was drawn in elusive snowy suggestions, rather than lines, against a green space of intense yet pale tint. The moon, now nearing the full, hung over the wooded valley, and aided the ice and the crust of snow to show its bleak, wan, wintry aspect; a tiny spark glowed in its depths from some open door of an isolated home. Over it all a mist was rising from the east, drawing its fleecy but opaque curtain. Already it had climbed the mountain-side and advanced, windless, soundless, overwhelming, annihilating all before and beneath it. The old hotel had disappeared, save that here and there a gaunt gable protruded and was withdrawn, showed once more, and once more was submerged.

A horse's head suddenly looking out of the enveloping mist close to his shoulder gave him the first intimation of the arrival, the secret silent waiting, of those whom he had directed hither. That the saddles were empty he saw a moment later. The animals stood together in a row, hitched to the rack. No disturbance sounded from the silent building. The event was in abeyance. The fugitive in hiding was doubtless at ease, unsuspecting, while the noiseless search of the officers for his quarters was under way.

With a thrill of excitement Keenan crept stealthily through an open passage and into the old grass-grown spaces of the quadrangle. Night possessed the place, but the cloud seemed denser than the darkness. He was somehow sensible of its convolutions as he stood against the wall and strained his eyes into the dusk. Suddenly it was penetrated by a milky-white glimmer, a glimmer duplicated at equidistant points, each fading as its successor sprang into brilliance. The next moment he understood its significance. It had come from the blurred windows of the old ball-room. Millicent had lighted her candle as she searched for the fugitive's quarters; she was passing down the length of the old house on the second story, and suddenly she emerged upon the gallery. She shielded the feeble flicker with her hand; her white-hooded head gleamed as with an aureola as the divergent rays rested on the opaque mist; and now and again she clutched the baluster and walked with tremulous care, for the flooring was rotten here and there, and ready to crumble away. Her face was pallid, troubled; and Dundas, who had been warned by the tramp of horses and the tread of men, and who had descended the stairs, revolver in hand, ready to slip away if he might under cover of the mist, paused appalled, gazing across the quadrangle as on an apparition—the sight so familiar to his senses, so strange to his experience. He saw in an abrupt shifting of the mist that there were other figures skulking in doorways, watching her progress. The next moment she leaned forward to clutch the baluster, and the light of the candle fell full on Emory Keenan, lurking in the open passage.

A sudden sharp cry of "Surrender!"

The young mountaineer, confused, swiftly drew his pistol. Others were swifter still. A sharp report rang out into the chill crisp air, rousing all the affrighted echoes — a few faltering steps, a heavy fall, and for a long time Emory Keenan's life-blood stained the floor of the promenade. Even when it had faded, the rustic gossips came often and gazed at the spot with morbid interest, until, a decade later, an enterprising proprietor removed the floor and altered the shape of that section of the building out of recognition.

The escape of Dundas was easily effected. The deputy sheriff, confronted with the problem of satisfactorily accounting for the death of a man who had committed no offence against public polity, was no longer formidable. His errand had been the arrest of a horse-thief, well-known to him, and he had no interest in pursuing a fugitive, however obnoxious to the law, whose personal description was so different from that of the object of his search.

Time restored to Dundas his former place in life and the esteem of his fellow-citizens. His stay in the mountains was an episode which he will not often recall, but sometimes volition fails, and he marvels at the strange fulfilment of the girl's vision; he winces to think that her solicitude for his safety should have cost her her lover; he wonders whether she yet lives, and whether that tender troubled phantom, on nights when the wind is still and the moon is low and the mists rise, again joins the strange, elusive, woful company crossing the quaking foot-bridge.

HIS "DAY IN COURT"

It had been a hard winter along the slopes of the Great Smoky Mountains, and still the towering treeless domes were covered with snow, and the vagrant winds were abroad, rioting among the clifty heights where they held their tryst, or raiding down into the sheltered depths of the Cove, where they seldom intruded. Nevertheless, on this turbulent rush was borne in the fair spring of the year. The fragrance of the budding wild-cherry was to be discerned amidst the keen slanting javelins of the rain. A cognition of the renewal and the expanding of the forces of nature pervaded the senses as distinctly as if one might hear the grass growing, or feel along the chill currents of the air the vernal pulses thrill. Night after night in the rifts of the breaking clouds close to the horizon was glimpsed the stately sidereal Virgo, prefiguring and promising the harvest, holding in her hand a gleaming ear of corn. But it was not the constellation which the tumultuous torrent at the mountain's base reflected in a starry glitter. From the hill-side above a light cast its broken image among the ripples, as it shone for an instant through the bosky laurel, white, stellular, splendid—only a tallow dip suddenly placed in the window of a log-cabin, and as suddenly withdrawn.

For a gruff voice within growled out a remonstrance : " What ye doin' that fur, Steve ?   Hev
that thar candle got enny call ter bide in that thar
winder ?"

The interior, contrary to the customary aspect of
the humble homes of the region, was in great dis-
array. Cooking utensils stood uncleaned about the
hearth ; dishes and bowls of earthen-ware were as-
sembled upon the table in such numbers as to suggest that several meals had been eaten without the
ceremony of laying the cloth anew, and that in default of washing the crockery it had been re-enforced
from the shelf so far as the limited store might admit.   Saddles and spinning-wheels, an ox-yoke
and trace-chains, reels and wash-tubs, were incongruously pushed together in the corners.   Only one
of the three men in the room made any effort
to reduce the confusion to order.   This was the
square-faced, black-bearded, thick-set young fellow
who took the candle from the window, and now advanced with it toward the hearth, holding it at an
angle that caused the flame to swiftly melt the tallow, which dripped generously upon the floor.

" I hev seen Eveliny do it," he said, excitedly
justifying himself.   " I noticed her sot the candle
in the winder jes' las' night arter supper."   He
glanced about uncertainly, and his patience seemed
to give way suddenly.   "Dad-burn the old candle!
I dunno *whar* ter set it," he cried, desperately, as
he flung it from him, and it fell upon the floor close
to the wall.

The dogs lifted their heads to look, and one soft-
stepping old hound got up with the nimbleness of

expectation, and, with a prescient gratitude astir
in his tail, went and sniffed at it. His aspect
drooped suddenly, and he looked around in reproach at Stephen Quimbey, as if suspecting a
practical joke. But there was no merriment in the
young mountaineer's face. He threw himself into
his chair with a heavy sigh, and desisted for a time
from the unaccustomed duty of clearing away the
dishes after supper.

"An' 'ain't ye got the gumption ter sense what
Eveliny sot the candle in the winder fur?" his
brother Timothy demanded, abruptly—"ez a sign
ter that thar durned Abs'lom Kittredge."

The other two men turned their heads and looked
at the speaker with a poignant intensity of interest.
"I 'lowed ez much when I seen that light ez I war
a-kemin' home las' night," he continued; "it shined
spang down the slope acrost the ruver an' through
all the laurel; it looked plumb like a star that hed
fell ter yearth in that pitch-black night. I dunno how I s'picioned it, but ez I stood thar an'
gazed I knowed somebody war a-standin' an' gazin'
too on the foot-bredge a mite ahead o' me. I
couldn't see him, an' he couldn't turn back an' pass
me, the bredge bein' too narrer. He war jes
obligated ter go on. I hearn him breathe quick;
then—pit-pat, pit-pat, ez he walked straight toward
that light. An' he be 'bleeged ter hev hearn me,
fur arter I crost I stopped. Nuthin'. Jes' a whisper o' wind, an' jes' a swishin' from the ruver. I
knowed then he hed turned off inter the laurel.
An' I went on, a-whistlin' ter make him 'low ez I
never s'picioned nuthin'. An' I kem inter the

5

house an' tole dad ez he'd better be a-lookin' arter
Eveliny, fur I b'lieved she war a-settin' her head
ter run away an' marry Abs'lom Kittredge."

"Waal, I ain't right up an' down sati'fied we
oughter done what we done," exclaimed Stephen,
fretfully. "It don't 'pear edzacly right fur three
men ter fire on one."

Old Joel Quimbey, in his arm-chair in the chimney-corner, suddenly lifted his head—a thin head
with fine white hair, short and sparse, upon it. His
thin, lined face was clear-cut, with a pointed chin
and an aquiline nose. He maintained an air of indignant and rebellious grief, and had hitherto sat
silent, a gnarled and knotted hand on either arm
of his chair. His eyes gleamed keenly from under his heavy brows as he turned his face upon
his sons. "How could we know thar warn't but
one, eh?"

He had not been a candidate for justice of the
peace for nothing; he had absorbed something of
the methods and spirit of the law through sheer
propinquity to the office. "We-uns wouldn't be
persumed ter *know.*" And he ungrudgingly gave
himself all the benefit of the doubt that the law
accords.

"That's a true word!" exclaimed Stephen, quick
to console his conscience. "Jes' look at the fac's,
now. We-uns in a plumb black midnight hear a man
a-gittin' over our fence; we git our rifles; a-peekin'
through the chinkin' we ketch a glimge o' him—"

"Ha!" cried out Timothy, with savage satisfaction, "we seen him by the light she set ter lead
him on!"

"OLD JOEL QUIMBEY"

.

He was tall and lank, with a delicately hooked
nose, high cheek-bones, fierce dark eyes, and dark
eyebrows, which were continually elevated, corru-
gating his forehead.   His hair was black, short and
straight, and he was clad in brown jeans, as were
the others, with great cowhide boots reaching to
the knee.   He fixed his fiery intent gaze on his
brother as the slower Stephen continued, "An' so
we blaze away—"

"An' one durned fool's so onlucky ez ter hit him
an' not kill him," growled Timothy, again interrupt-
ing.   "An' so whilst Eveliny runs out a-screamin',
'He's dead! he's dead!—ye hev shot him dead!'
we-uns make no doubt but he *is* dead, an' load up
agin, lest his frien's mought rush in on we-uns
whilst we hedn't no use o' our shootin'-irons.   An'
suddint—ye can't hear nuthin' but jes' a owel hoot-
in' in the woods, or old Pa'son Bates's dogs a-howlin'
acrost the Cove.   An' we go out with a lantern,
an' thar's jes' a pool o' blood in the dooryard, an'
bloody tracks down ter the laurel."

"Eveliny gone!" cried the old man, smiting his
hands together; "my leetle darter!   The only one
ez never gin me enny trouble.   I couldn't hev
made out ter put up with this hyar worl' no longer
when my wife died ef it hedn't been fur Eveliny.
Boys war wild an' mischeevious, an' folks outside
don't keer nuthin' 'bout ye—ef they *war* ter 'lect
ye ter office 'twould be ter keep some other feller
from hevin' it, 'kase they 'spise him more'n ye.
An' hyar she's runned off an' married old Tom
Kittredge's gran'son, Josiah Kittredge's son—when
our folks 'ain't spoke ter none o' 'em fur fifty year

—Josiah Kittredge's son — ha ! ha ! ha !" He laughed aloud in tuneless scorn of himself and of this freak of froward destiny and then fell to wringing his hands and calling upon Evelina.

The flare from the great chimney-place genially played over the huddled confusion of the room and the brown logs of the wall, where the gigantic shadows of the three men mimicked their every gesture with grotesque exaggeration. The rainbow yarn on the warping bars, the strings of red-pepper hanging from the ceiling, the burnished metallic flash from the guns on their racks of deer antlers, served as incidents in the monotony of the alternate yellow flicker and brown shadow. Deep under the blaze the red coals pulsated, and in the farthest vistas of the fire quivered a white heat.

"Old Tom Kittredge," the father resumed, after a time, "he jes' branded yer gran'dad's cattle with his mark; he jes' cheated yer gran'dad, my dad, out'n six head o' cattle."

" But then," said the warlike Timothy, not willing to lose sight of reprisal even in vague reminiscence, "he hed only one hand ter rob with arter that, fur I hev hearn ez how when gran'dad got through with him the doctor hed ter take his arm off."

" Sartainly, sartainly," admitted the old man, in quiet assent. "An' Josiah Kittredge he put out the eyes of a horse critter o' mine right thar at the court-house door—"

" Waal, arterward, we - uns fired his house over his head," put in Tim.

" An' Josiah Kittredge an' me," the old man went on, "we-uns clinched every time we met in

this mortal life. Every time I go past the grave-
yard whar he be buried I kin feel his fingers on
my throat. He had a nervy grip, but no variation;
he always tuk holt the same way."

"'Pears like ter me ez 'twar a fust-rate time ter
fetch out the rifles again," remarked Tim, "this
mornin', when old Pa'son Bates kem up hyar an'
'lowed ez he hed married Eveliny ter Abs'lom
Kittredge on his death-bed; 'So be, pa'son,' I say.
An' he tuk off his hat an' say, 'Thank the Lord,
this will heal the breach an' make ye frien's!' An'
I say, 'Edzacly, pa'son, ef it *air* Abs'lom's death-
bed; but them Kittredges air so smilin' an' deceivin' I be powerful feared he'll cheat the King o'
Terrors himself. I'll forgive 'em ennything—*over
his grave.*' "

"Pa'son war tuk toler'ble suddint in his temper,"
said the literal Steve. "I hearn him call yer talk
onchristian, cussed sentiments, ez he put out."

"Ye mus' keep up a Christian sperit, boys; that's
the main thing," said the old man, who was esteemed
very religious, and a pious Mentor in his own family. He gazed meditatively into the fire. "What
ailed Eveliny ter git so tuk up with this hyar Abs'-
lom? What made her like him?" he propounded.

"His big eyes, edzacly like a buck's, an' his
long yaller hair," sneered the discerning Timothy,
with the valid scorn of a big ugly man for a slim
pretty one. "'Twar jes 'count o' his long yaller
hair his mother called him Abs'lom. He war
named Pete or Bob, I disremember what—suthin'
common—till his hair got so long an' curly, an' he
sot out ter be so plumb all-fired beautiful, an' his

mother named him agin; this time Abs'lom, arter
the king's son, 'count o' his yaller hair."

"Git hung by his hair some o' these days in the
woods, like him the Bible tells about; that happened
ter the sure-enough Abs'lom," suggested Stephen,
hopefully.

"Naw, sir," said Tim; "when Abs'lom Kittredge
gits hung it 'll be with suthin' stronger'n hair; he'll
stretch hemp." He exchanged a glance of triumphant prediction with his brother, and anon
gazed ruefully into the fire.

"Ye talk like ez ef he war goin' ter live, boys,"
said old Joel Quimbey, irritably. "Pa'son 'lowed
he war powerful low."

"Pa'son said he'd never hev got home alive
'thout she'd holped him," said Stephen. "She jes'
tuk him an' drug him plumb ter the bars, though I
don't see how she done it, slim leetle critter ez she
be; an' thar she holped him git on his beastis; an'
then — I declar' I feel ez ef I could kill her fur
a-demeanin' of herself so—she led that thar horse,
him a-ridin' an' a-leanin' on the neck o' the beastis,
two mile up the mountain, through the night."

"Waal, let her bide thar. I'll look on her face
no mo'," declared the old man, his toothless jaw
shaking. "Kittredge she be now, an' none o' the
name kin come a-nigh me. How be I ever a-goin'
'bout 'mongst the folks at the settlement agin with
my darter married ter a Kittredge? How Josiah
an' his dad mus' be a-grinnin' in thar graves at me
this night! An' I 'low they hev got suthin' ter grin
about."

And suddenly his grim face relaxed, and once

more he began to smite his hands together and to
call aloud for Evelina.

Timothy could offer no consolation, but stared
dismally into the fire, and Stephen rose with a sigh
and addressed himself to pushing the spinning-
wheels and tubs and tables into the opposite corner of the room, in the hope of solving the enigma
of its wonted order.

It seemed to Evelina afterward that when she
climbed the rugged ways of the mountain slope in
that momentous night she left forever in the depths
of the Cove that free and careless young identity
which she had been. She did not accurately discriminate the moment in which she began to realize
that she was among her hereditary enemies, encompassed by a hatred nourished to full proportions and to a savage strength long before she drew
her first breath. The fact only gradually claimed
its share in her consciousness as the tension of anxiety for Absalom's sake relaxed, for the young
mountaineer's strength and vitality were promptly
reasserted, and he rallied from the wound and his
pallid and forlorn estate with the recuperative
power of the primitive man. By degrees she came
to expect the covert unfriendly glances his brother
cast upon her, the lowering averted mien oi her
sister-in-law, and now and again she surprised a
long, lingering, curious gaze in his mother's eyes.
They were all Kittredges! And she wondered how
she could ever have dreamed that she might live
happily among them—one of them; for her name
was theirs. And then perhaps the young husband

would stroll languidly in, with his long hair curling on his blue jeans coat-collar, and an assured smile in his dark brown eyes, and some lazy jest on his lips, certain of a welcoming laugh, for he had been so near to death that they all had a sense of acquisition in that he had been led back. For his sake they had said little ; his mother would busy herself in brewing his "yerb" tea, and his brother would offer to saddle the mare if he felt that he could ride, and they would all be very friendly together ; and his alien wife would presently slip out unnoticed into the "gyarden spot," where the rows of vegetables grew as they did in the Cove, turning upon her the same neighborly looks they wore of yore, and showing not a strange leaf among them. The sunshine wrapped itself in its old fine gilded gossamer haze and drowsed upon the verdant slopes; the green jewelled " Juny-bugs " whirred in the soft air ; the mould was as richly brown as in Joel Quimbey's own enclosure ; the flag-lilies bloomed beside the onion bed ; and the woolly green leaves of the sage wore their old delicate tint and gave out a familiar odor.

Among this quaint company of the garden borders she spent much of her time, now hoeing in a desultory fashion, now leaning on the long handle of the implement and looking away upon the far reaches of the purple mountains. As they stretched to vague distances they became blue, and farther on the great azure domes merged into a still more tender hue, and this in turn melted into a soft indeterminate tint that embellished the faint horizon. Her dreaming eyes would grow bright and wistful ;

her rich brown curling hair, set free by the yellow
sun-bonnet that slipped off her head and upon her
shoulders, would airily float backward in the wind;
there was a lithe grace in the slender figure, albeit
clad in a yellow homespun of a deep dye, and the
faded purplish neckerchief was caught about a
throat fairer even than the fair face, which was
delicately flushed. Absalom's mother, standing beside Peter, the eldest son, in the doorway, watched
her long one day.

"It all kem about from that thar bran dance,"
said Peter, a homely man, with a sterling, narrow-
minded wife and an ascetic sense of religion. "Thar
Satan waits, an' he gits nimbler every time ye shake
yer foot. The fiddler gin out the figger ter change
partners, an' this hyar gal war dancin' opposite Abs'-
lom, ez hed never looked nigh her till that day. The
gal didn't know *what* ter do; she jes' stood still;
but Abs'lom he jes' danced up ter her ez keerless
an' gay ez he always war, jes' like she war ennybody
else, an' when he held out his han' she gin him
hern, all a-trembly, an' lookin' up at him, plumb
skeered ter death, her eyes all wide an' sorter wish-
ful, like some wild thing trapped in the woods. An'
then the durned fiddler, moved by the devil, I'll be
be bound, plumb furgot ter change 'em back. So
they danced haff'n the day tergether. An' arter
that they war forever a-stealin' off an' accidentally
meetin' at the spring, an' whenst he war a-huntin'
or she drivin' up the cow, an' a-courtin' ginerally,
till they war promised ter marry."

"'Twarn't the bran dance; 'twar suthin' ez fleet-
in' an' ez useless," said his mother, standing in the

door and gazing at the unconscious girl, who was
leaning upon the hoe, half in the shadow of the
blooming laurel that crowded about the enclosure
and bent over the rail fence, and half in the burnished sunshine ; " she's plumb beautiful—thar's the
snare ez tangled Abs'lom's steps.   I never 'lowed
ter see the day ez could show enny comfort fur his
dad bein' dead, but we hev been spared some o' the
tallest cavortin' that ever war seen sence the Big
Smoky war built. Sometimes it plumb skeers me ter
think ez we-uns hev got a Quimbey abidin' up hyar
along o' we-uns in *his* house an' a-callin' o' herse'f
Kittredge.   I looks ter see him a-stalkin' roun' hyar
some night, too outdone an' aggervated ter rest in
his grave."

But the nights continued spectreless and peaceful on the Great Smoky, and the same serene stars
shone above the mountain as over the Cove.   Evelina could watch here, as often before, the rising
moon ascending through a rugged gap in the range,
suffusing the dusky purple slopes and the black
crags on either hand with a pensive glamour, and
revealing the river below by the amber reflection its
light evoked.   She often sat on the step of the
porch, her elbow on her knees, her chin in her
hand, following with her shining eyes the pearly
white mists loitering among the ranges.   Hear ! a
dog barks in the Cove, a cock crows, a horn is
wound, far, far away; it echoes faintly.   And once
more only the sounds of the night—that vague stir
in the windless woods, as if the forest breathes, the
far-away tinkle of water hidden in the darkness—
and the moon is among the summits.

The men remained within, for Absalom avoided
the chill night air, and crouched over the smouldering fire. Peter's wife sedulously held aloof from
the ostracized Quimbey woman. But her mother-
in-law had fallen into the habit of sitting upon
the porch these moonlit nights. The sparse, new-
ly-leafed hop and gourd vines clambering to its
roof were all delicately imaged on the floor, and
the old woman's clumsy figure, her grotesque sun-
bonnet, her awkward arm-chair, were faithfully re-
produced in her shadow on the log wall of the
cabin—even to the up-curling smoke from her pipe.
Once she suddenly took the stem from her mouth.
"Eveliny," she said, "'pears like ter me ye talk
mighty little. Thar ain't no use in gittin' tongue-
tied up hyar on the mounting."

Evelina started and raised her eyes, dilated with
a stare of amazement at this unexpected overture.

"I ain't keerin'," said the old woman, recklessly,
to herself, although consciously recreant to the traditions of the family, and sacrificing with a pang
her distorted sense of loyalty and duty to her
kindlier impulse. "I warn't born a Kittredge
nohow."

"Yes, 'm," said Evelina, meekly; "but I don't
feel much like talkin' noways; I never talked
much, bein' nobody but men-folks ter our house.
I'd ruther hear ye talk 'n talk myself."

"Listen at ye now! The headin' young folks o'
this kentry 'll never rest till they make thar elders
shoulder all the burdens. An' what air ye wantin' a
pore ole 'oman like me ter talk about?"

Evelina hesitated a moment, then looked up,

with a face radiant in the moonbeams. "Tell all
'bout Abs'lom—afore I ever seen him."

His mother laughed. "Ye air a powerful fool,
Eveliny."

The girl laughed a little, too. "I dunno ez I
want ter be no wiser," she said.

But one was his wife, and the other was his
mother, and as they talked of him daily and long,
the bond between them was complete.

"I hev got 'em both plumb fooled," the handsome Absalom boasted at the settlement, when the
gossips wondered once more, as they had often
done, that there should be such unity of interest
between old Joel Quimbey's daughter and old Josiah Kittredge's widow. As time went on many
rumors of great peace on the mountain-side came
to the father's ears, and he grew more testy daily
as he grew visibly older. These rumors multiplied
with the discovery that they were as wormwood and
gall to him. Not that he wished his daughter to
be unhappy, but the joy which was his grief and
humiliation was needlessly flaunted into his face;
the idlers about the county town had invariably a
new budget of details, being supplied, somewhat
maliciously, it must be confessed, by the Kittredges
themselves. The ceremony of planting one foot
on the neck of the vanquished was in their minds
one of the essential concomitants of victory. The
bold Absalom, not thoroughly known to either of the
women who adored him, was ingenious in expedi-
ents, and had applied the knowledge gleaned from
his wife's reminiscences of her home, her father,

and her brothers to more accurately aim his darts. Sometimes old Quimbey would fairly flee the town, and betake himself in a towering rage to his deserted hearth, to brood futilely over the ashes, and devise impotent schemes of vengeance.

He often wondered afterward in dreary retro-. spection how he had survived that first troublous year after his daughter's elopement, when he was so lonely, so heavy-hearted at home, so harried and angered abroad. His comforts, it is true, were amply insured : a widowed sister had come to preside over his household—a deaf old woman, who had much to be thankful for in her infirmity, for Joel Quimbey in his youth, before he acquired religion, had been known as a singularly profane man —"a mos' survigrus cusser"—and something of his old proficiency had returned to him. Perhaps public sympathy for his troubles strengthened his hold upon the regard of the community. For it was in the second year of Evelina's marriage, in the splendid midsummer, when all the gifts of nature climax to a gorgeous perfection, and candidates become incumbents, that he unexpectedly attained the great ambition of his life. He was said to have made the race for justice of the peace from sheer force of habit, but by some unexplained freak of popularity the oft-defeated candidate was successful by a large majority at the August election.

"Laws-a-massy, boys," he said, tremulously, to his triumphant sons, when the result was announced, the excited flush on his thin old face suffusing his hollow veinous temples, and rising into his fine white hair, "how glad Eveliny would hev been ef—

ef—" He was about to say if she had lived, for
he often spoke of her as if she were dead. He
turned suddenly back, and began to eagerly absorb
the details of the race, as if he had often before
been elected, with calm superiority canvassing the
relative strength, or rather the relative weakness, of
the defeated aspirants.

He could scarcely have measured the joy which
the news gave to Evelina. She was eminently sus-
ceptible of the elation of pride, the fervid glow of
success; but her tender heart melted in sympathetic divination of all that this was to him who
had sought it so long, and so unabashed by defeat.
She pined to see his triumph in his eyes, to hear it
in his voice. She wondered—nay, she knew that
he longed to tell it to her. As the year rolled
around again to summer, and she heard from time
to time of his quarterly visits to the town as a member of the worshipful Quarterly County Court, she
began to hope that, softened by his prosperity,
lifted so high by his honors above all the cavillings
of the Kittredges, he might be more leniently disposed toward her, might pity her, might even go so
far as to forgive.

But none of her filial messages reached her father's fiery old heart.

"Ye'll be sure, Abs'lom, ef ye see Joe Boyd in
town, ye'll tell him ter gin dad my respec's, an' the
word ez how the baby air a-thrivin', an' I wants ter
fotch him ter see the fambly at home, ef they'll
lemme."

Then she would watch Absalom with all the
confidence of happy anticipation, as he rode off

down the mountain with his hair flaunting, and his
spurs jingling, and his shy young horse curveting.

But no word ever came in response; and sometimes she would take the child in her arms and
carry him down a path, worn smooth by her own
feet, to a jagged shoulder thrust out by the mountain where all the slopes fell away, and a crag beetled
over the depths of the Cove. Thence she could
discern certain vague lines marking the enclosure,
and a tiny cluster of foliage hardly recognizable
as the orchard, in the midst of which the cabin
nestled. She could not distinguish them, but she
knew that the cows were coming to be milked, lowing and clanking their bells tunefully, fording the
river that had the sunset emblazoned upon it, or
standing flank deep amidst its ripples; the chickens
might be going to roost among the althea bushes;
the lazy old dogs were astir on the porch. She
could picture her brothers at work about the barn;
most often a white-haired man who walked with a
stick—alack! she did not fancy how feebly, nor
that his white hair had grown long and venerable,
and tossed in the breeze. "Ef he would jes lemme
kem fur one haff'n hour!" she would cry.

But all her griefs were bewept on the crag, that
there might be no tears to distress the tender-
hearted Absalom when she should return to the
house.

The election of Squire Quimbey was a sad blow
to the arrogant spirit of the Kittredges. They had
easily accustomed themselves to ascendency, and
they hotly resented the fact that fate had forborne
the opportunity to hit Joel Quimbey when he was

down. They had used their utmost influence to defeat him in the race, and had openly avowed their desire to see him bite the dust. The inimical feeling between the families culminated one rainy autumnal day in the town where the quarterly county court was in session.

A fire had been kindled in the great rusty stove, and crackled away with grudging merriment inside, imparting no sentiment of cheer to the gaunt bare room, with its dusty window-panes streaked with rain, its shutters drearily flapping in the wind, and the floor bearing the imprint of many boots burdened with the red clay of the region. The sound of slow strolling feet in the brick-paved hall was monotonous and somnolent.

Squire Quimbey sat in his place among the justices. Despite his pride of office, he had not the heart for business that might formerly have been his. More than once his attention wandered. He looked absently out of the nearest window at the neighboring dwelling — a little frame-house with a green yard; a well-sweep was defined against the gray sky, and about the curb a file of geese followed with swaying gait the wise old gander. "What a hand for fow-*els* Eveliny war!" he muttered to himself; "an' she hed luck with sech critters." He used the obituary tense, for Evelina had in some sort passed away.

He rubbed his hand across his corrugated brow, and suddenly he became aware that her husband was in the room, speaking to the chairman of the county court, and claiming a certificate in the sum of two dollars each for the scalps of one wolf, "an' one painter," he continued, laying the small furry repulsive objects upon the desk, "an' one dollar fur the skelp of one wild-cat." He was ready to take his oath that these animals were killed by him running at large in this county.

He had stooped a little in making the transfer. He came suddenly to his full height, and stood with one hand in his leather belt, the other shouldering his rifle. The old man scanned him curiously. The crude light from the long windows was full upon his tall slim figure; his yellow hair curled down upon the collar of his blue jeans coat; his great miry boots were drawn high over the trousers to the knee; his pensive deer-like eyes brightened with a touch of arrogance and enmity as, turning slowly to see who was present, his glance encountered his father-in-law's fiery gaze.

"Mr. Cheerman! Mr. Cheerman!" exclaimed the old man, tremulously, "lemme examinate that thar wild-cat skelp. Thanky, sir; thanky, sir; I wanter see ef 'tain't off'n the head o' some old tame tom-cat. An' this air a painter's"—affecting to scan it by the window—"two ears 'cordin' to law; yes, sir, two; and this"—his keen old face had all the white light of the sad gray day on its bleaching hair and its many lines, and his eager old hands trembled with the excitement of the significant satire he enacted—"an' this air a wolf's, ye say? Yes; it's a Kittredge's; same thing, Mr. Cheerman, by a diff'ent name; nuthin' in the code 'bout'n a premium fur a Kittredge's skelp; but same natur'; coward, bully, thief—*thief!*"

The words in the high cracked voice rang from

6

the bare walls and bare floors as he tossed the scalps trom him, and sat down, laughing silently in painful, mirthless fashion, his toothless jaw quivering, and his shaking hands groping for the arms of his chair.

"Who says a Kittredge air a thief says a lie!" cried out the young man, recovering from his tense surprise. "I don't keer how old he be," he stipulated—for he had not thought to see her father so aged—"he lies."

The old man fixed him with a steady gaze and a sudden alternation of calmness. "Ye air a Kittredge; ye stole my daughter from me."

"I never. She kem of her own accord."

"Damn ye!" the old man retorted to the unwelcome truth. There was nothing else for him to say. "Damn the whole tribe of ye; everything that goes by the accursed name of Kittredge, that's got a drop o' yer blood, or a bone o' yer bones, or a puff o' yer breath—"

"Squair! squair!" interposed an officious old colleague, taking him by the elbow, "jes' quiet down now; ye air a-cussin' yer own gran'son."

"So be! so be!" cried the old man, in a frenzy of rage. "Damn 'em all—all the Kittredge tribe!" He gasped for breath; his lips still moved speechlessly as he fell back in his chair.

Kittredge let his gun slip from his shoulder, the butt ringing heavily as it struck upon the floor. "I ain't a-goin' ter take sech ez that off'n ye, old man," he cried, pallid with fury, for be it remembered this grandson was that august institution, a first baby. "He sha'n't sit up thar an' cuss the

baby, Mr. Cheerman." He appealed to the pre-
siding justice, holding up his right arm as tremulous as old Quimbey's own. "I want the law! I
ain't a-goin' ter tech a old man like him, an' my
wife's father, so I ax in the name o' peace fur the
law. Don't deny it"—with a warning glance—
"'kase I ain't school-larned, an' dunno how ter
get it. Don't ye deny me the law! I *know* the
law don't 'low a magistrate an' a jestice ter cuss in
his high office, in the presence of the county court.
I want the law! I want the law!"

The chairman of the court, who had risen in his
excitement, turning eagerly first to one and then to
the other of the speakers, striving to silence the
colloquy, and in the sudden surprise of it at a momentary loss how to take action, sat down abruptly, and with a face of consternation. Profanity
seemed to him so usual and necessary an incident
of conversation that it had never occurred to him
until this moment that by some strange aberration
from the rational estimate of essentials it was entered in the code as a violation of law. He would
fain have overlooked it, but the room was crowded
with spectators. The chairman would be a candidate for re-election as justice of the peace at the
expiration of his term. And after all what was old
Quimbey to him, or he to old Quimbey, that, with
practically the whole town looking on, he should
destroy his political prospects and disregard the
dignity of his office. He had a certain twinge of
conscience, and a recollection of the choice and
fluent oaths of his own repertory, but as he turned
over the pages of the code in search of the section

he deftly argued that they were uttered in his own presence as a person, not as a justice.

And so for the first time old Joel Quimbey appeared as a law-breaker, and was duly fined by the worshipful county court fifty cents for each oath, that being the price at which the State rates the expensive and impious luxury of swearing in the hearing of a justice of the peace, and which in its discretion the court saw fit to adopt in this instance.

The old man offered no remonstrance; he said not a word in his own defence. He silently drew out his worn wallet, with much contortion of his thin old anatomy in getting to his pocket, and paid his fines on the spot. Absalom had already left the room, the clerk having made out the certificates, the chairman of the court casting the scalps into the open door of the stove, that they might be consumed by fire according to law.

The young mountaineer wore a heavy frown, and his heart was ill at ease. He sought some satisfaction in the evident opinion of the crowd which now streamed out, for the excitements within were over, that he had done a fine thing; a very clever thought, they considered it, to demand the law of Mr. Chairman, that one of their worships should be dragged from the bench and arraigned before the quarterly county court of which he was a member. The result gave general satisfaction, although there were those who found fault with the court's moderation, and complained that the least possible cognizance had been taken of the offence.

"Ho! ho! ho!" laughed an old codger in the street. "I jes knowed that hurt old Joel Quimbey

wuss 'n ef a body hed druv a knife through him;
he's been so proud o' bein' jestice 'mongst his bet-
ters, an' bein' 'lected at las', many times ez he hev
run.   Waal, Abs'lom, ye hev proved thar's law fur
jestices too.   I tell ye ye hev got sense in yer skull-
i-bone."

But Absalom hung his head before these con-
gratulations; he found no relish in the old man's
humbled pride.   Yet had he not cursed the baby,
lumping him among the Kittredges?   Absalom
went about for a time, with a hopeful anxiety in
his eyes, searching for one of the younger Quimbeys, in order to involve him in a fight that might
have a provocation and a result more to his mind.
Somehow the recollection of the quivering and aged
figure of his wife's father, of the smitten look on his
old face, of his abashed and humbled demeanor
before the court, was a reproach to him, vivid and
continuously present with his repetitious thoughts
forever re - enacting the scene.   His hands trembled ; he wanted to lay hold on a younger man, to
replace this æsthetic revenge with a quarrel more
wholesome in the estimation of his own conscience.
But the Quimbey sons were not in town to-day.
He could only stroll about and hear himself praised
for this thing that he had done, and wonder how
he should meet Evelina with his conscience thus ar-
rayed against himself for her father's sake.   "Plumb
turned Quimbey, I swear," he said, in helpless reproach to this independent and coercive moral force
within.   His dejection, he supposed, had reached
its lowest limits, when a rumor pervaded the town,
so wild that he thought it could be only fantasy.

It proved to be fact.   Joel Quimbey, aggrieved,
humbled, and indignant, had resigned his office, and
as Absalom rode out of town toward the mountains,
he saw the old man in his crumpled brown jeans
suit, mounted on his white mare, jogging down the
red clay road, his head bowed before the slanting
lines of rain, on his way to his cheerless fireside.
He turned off presently, for the road to the levels
of the Cove was not the shorter cut that Absalom
travelled to the mountains.   But all the way the
young man fancied that he saw from time to time,
as the bridle-path curved in the intricacies of the
laurel, the bowed old figure among the mists, jogging along, his proud head and his stiff neck bent to
the slanting rain and the buffets of his unkind fate.
And yet, pressing the young horse to overtake him,
Absalom could find naught but the fleecy mists
drifting down the bridle-path as the wind might
will, or lurking in the darkling nooks of the laurel
when the wind would.

The sun was shining on the mountains, and Absalom went up from the sad gray rain and through
the gloomy clouds of autumn hanging over the
Cove into a soft brilliant upper atmosphere — a
generous after-thought of summer—and the warm
brightness of Evelina's smile.   She stood in the
doorway as she saw him dismounting, with her finger on her lips, for the baby was sleeping: he put '
much of his time into that occupation.   The tiny
gourds hung yellow among the vines that clam-
bered over the roof of the porch, and a brave jack-
bean—a friend of the sheltering eaves—made shift

to bloom purple and white, though others of the kind hung crisp and sere, and rattled their dry bones in every gust. The "gyarden spot" at the side of the house was full of brown and withered skeletons of the summer growths; among the crisp blades of the Indian-corn a sibilant voice was forever whispering; down the tawny-colored vistas the pumpkins glowed. The sky was blue; the yellow hickory flaming against it and hanging over the roof of the cabin was a fine color to see. The red sour-wood tree in the fence corner shook out a myriad of white tassels; the rolling tumult of the gray clouds below thickened, and he could hear the rain a-falling—falling into the dreary depths of the Cove.

All this for him: why should he disquiet himself for the storm that burst upon others?

Evelina seemed a part of the brightness; her dark eyes so softly alight, her curving red lips, the faint flush in her cheeks, her rich brown hair, and the purplish kerchief about the neck of her yellow dress. Once more she looked smilingly at him, and shook her head and laid her finger on her lip.

" I oughter been sati'fied with all I got, stiddier hectorin' other folks till they 'ain't got no heart ter hold on ter what they been at sech trouble ter git," he said, as he turned out the horse and strode gloomily toward the house with the saddle over his arm.

" Hev ennybody been spiteful ter you-uns terday?" she asked, in an almost maternal solicitude, and with a flash of partisan anger in her eyes.

" Git out'n my road, Eveliny," he said, fretfully,

pushing by, and throwing the saddle on the floor. There was no one in the room but the occupant of the rude box on rockers which served as cradle.

Absalom had a swift, prescient fear. "She'll git it all out'n me ef I don't look sharp," he said to himself. Then aloud, "Whar's mam?" he demanded, flinging himself into a chair and looking loweringly about.

"Topknot hev jes kem off'n her nest with fourteen deedies, an' she an' 'Melia hev gone ter the barn ter see 'bout'n 'em."

"Whar's Pete?"

"A-huntin'."

A pause. The fire smouldered audibly; a hickory-nut fell with a sharp thwack on the clapboards of the roof, and rolled down and bounded to the ground.

Suddenly: "I seen yer dad ter-day," he began, without coercion. "He gin me a cussin', in the courtroom, 'fore all the folks. He cussed all the Kittredges, *all* o' 'em; him too"—he glanced in the direction of the cradle—"cussed 'em black an' blue, an' called me a *thief* fur marryin' ye an kerryin' ye off."

Her face turned scarlet, then pale. She sat down, her trembling hands reaching out to rock the cradle, as if the youthful Kittredge might be disturbed by the malediction hurled upon his tribe. But he slept sturdily on.

"Waal, now," she said, making a great effort at self-control, "ye oughtn't ter mind it. Ye know he war powerful tried. I never purtended ter be ez sweet an' pritty ez the baby air, but how would you-uns feel ef somebody ye despised war ter kem hyar an' tote him off from we-uns forever?"

"I'd cut thar hearts out," he said, with prompt barbarity.

"Thar, now!" exclaimed his wife, in triumphant logic.

He gloomily eyed the smouldering coals. He was beginning to understand the paternal sentiment. By his own heart he was learning the heart of his wife's father.

"I'd chop 'em inter minch-meat," he continued, carrying his just reprisals a step further.

"Waal, don't do it right now," said his wife, trying to laugh, yet vaguely frightened by his vehemence.

"Eveliny," he cried, springing to his feet, "I be a-goin' ter tell ye all 'bout'n it. I jes called on the cheerman fur the law agin him."

"Agin *dad!*—the law!" Her voice dropped as she contemplated aghast this terrible uncomprehended force brought to oppress old Joel Quimbey; she felt a sudden poignant pang for his forlorn and lonely estate.

"Never mind, never mind, Eveliny," Absalom said, hastily, repenting of his frantic candor and seeking to soothe her.

"I *will* mind," she said, sternly. "What hev ye done ter dad?"

"Nuthin'," he replied, sulkily—"nuthin'."

"Ye needn't try ter fool me, Abs'lom Kittredge. Ef ye ain't minded ter tell me, I'll foot it down ter town an' find out. What did the law do ter him?"

"Jes fined him," he said, striving to make light of it.

"An' ye done that fur—*spite!*" she cried. "A-settin' the law ter chouse a old man out'n money, fur gittin' mad an' sayin' ye stole his only darter. Oh, I'll answer fur him"—she too had risen; her hand trembled on the back of the chair, but her face was scornfully smiling—"he don't mind the *money;* he'll never git you-uns' *fined* ter pay back the gredge. He don't take his wrath out on folkses' *wallets;* he grips thar throats, or teches the trigger o' his rifle. Laws-a-massy! takin' out yer gredge that-a-way! It's *ye* poorer fur them dollars, Abs'lom —'tain't him." She laughed satirically, and turned to rock the cradle.

"What d'ye want me ter do? Fight a old man?" he exclaimed, angrily.

She kept silence, only looking at him with a flushed cheek and a scornful laughing eye.

He went on, resentfully: "I ain't 'shamed," he stoutly asserted. "Nobody 'lowed I oughter be. It's him, plumb bowed down with shame."

"The shoe's on the t'other foot," she cried. "It's ye that oughter be 'shamed, an' ef ye ain't, it's more shame ter ye. What hev he got ter be 'shamed of?"

"'Kase," he retorted, "he war fetched up afore a court on a crim'nal offence—a-cussin' afore the court! Ye may think it's no shame, but he do; he war so 'shamed he gin up his office ez jestice o' the peace, what he hev run fur four or five times, an' always got beat 'ceptin' wunst."

"Dad!" but for the whisper she seemed turning

to stone; her dilated eyes were fixed as she stared into his face.

"An' I seen him a-ridin' off from town in. the rain arterward, his head hangin' plumb down ter the saddle-bow."

Her amazed eyes were still fastened upon his face, but her hand no longer trembled on the back of the chair.

He suddenly held out his own hand to her, his sympathy and regret returning as he recalled the picture of the lonely wayfarer in the rain that had touched him so. "Oh, Eveliny!" he cried, "I never war so beset an' sorry an'—"

She struck his hand down; her eyes blazed. Her aspect was all instinct with anger.

"I do declar' I'll never furgive ye—ter spite him so—an' kem an' tell *me!* An' shame him so ez he can't hold his place—an' kem an' tell *me!* An' bow him down so ez he can't show his face whar he hev been so respected by all—an' kem an' tell *me!* An' all fur spite, fur he hev got nuthin' ye want now. An' I gin him up an' lef' him lonely, an' all fur you-uns. Ye air mean, Abs'lom Kittredge, an' I'm the mos' fursaken fool on the face o' the yearth!"

He tried to speak, but she held up her hand in expostulation.

"Nare word—fur I won't answer. I do declar' I'll never speak ter ye agin ez long ez I live."

He flung away with a laugh and a jeer. "That's right," he said, encouragingly; "plenty o' men would be powerful glad ef thar wives would take pattern by that."

He caught up his hat and strode out of the room. He busied himself in stabling his horse, and in looking after the stock. He could hear the women's voices from the loft of the barn as they disputed about the best methods of tending the newly hatched chickens, that had chipped the shell so late in the fall as to be embarrassed by the frosts and the coming cold weather. The last bee had ceased to drone about the great crimson prince's-feather by the door-step, worn purplish through long flaunting, and gone to seed. The clouds were creeping up and up the slope, and others were journeying hither from over the mountains. A sense of moisture was in the air, although a great column of dust sprang up from the dry corn-field, with panic-stricken suggestions, and went whirling away, carrying off withered blades in the rush. The first drops of rain were pattering, with a resonant timbre in the midst, when Pete came home with a newly killed deer on his horse, and the women, with fluttering skirts and sun-bonnets, ran swiftly across from the barn to the back door of the shed-room. Then the heavy downpour made the cabin rock.

"Why, Eveliny an' the baby oughtn't ter be out in this hyar rain — they'll be drenched," said the old woman, when they were all safely housed except the two. "Whar be she?"

"A-foolin' in the gyarden spot a-getherin' seed an' sech, like she always be," said the sister-in-law, tartly.

Absalom ran out into the rain without his hat, his heart in the clutch of a prescient terror. No; the summer was over for the garden as well as for him; all forlorn and rifled, its few swaying shrubs

tossed wildly about, a mockery of the grace and
bloom that had once embellished it.   His wet hair
streaming backward in the wind caught on the
laurel boughs as he went down and down the tangled path that her homesick feet had worn to the
crag which overlooked the Cove.   Not there!   He
stood, himself enveloped in the mist, and gazed
blankly into the folds of the dun-colored clouds
that with tumultuous involutions surged above the
valley and baffled his vision.   He realized it with
a sinking heart.   She was gone.

That afternoon—it was close upon nightfall—
Stephen Quimbey, letting down the bars for the
cows, noticed through the slanting lines of rain,
serried against the masses of sober-hued vapors
which hid the great mountain towering above the
Cove, a woman crossing the foot-bridge.   He
turned and lifted down another bar, and then
looked again.   Something was familiar in her aspect, certainly.   He stood gravely staring.   Her
sun-bonnet had fallen back upon her shoulders,
and was hanging loosely there by the strings tied
beneath her chin; her brown hair, dishevelled by
the storm, tossed back and forth in heavy wave-
less locks, wet through and through.   When the
wind freshened they lashed, thong-like, her pallid
oval face; more than once she put up her hand
and tried to gather them together, or to press them
back—only one hand, for she clasped a heavy bundle in her arms, and as she toiled along slowly up
the rocky slope, Stephen suddenly held his palm
above his eyes.   The recognition was becoming

definite, and yet he could scarcely believe his
senses: was it indeed Evelina, wind-tossed, tempest-
beaten, and with as many tears as rain-drops on her
pale cheek?   Evelina, forlorn and sorry, and with
swollen sad dark eyes, and listless exhausted step
—here again at the bars, where she had not stood
since she dragged her wounded lover thence on
that eventful night two years and more ago.

Resentment for the domestic treachery was up-
permost in his mind, and he demanded surlily,
when she had advanced within the sound of his
words, "What hev ye kem hyar fur?"

"Ter stay," she responded, briefly.

His hand in an uncertain gesture laid hold upon
his tuft of beard.

"Fur good?" he faltered, amazed.

She nodded silently.

He stooped to lift down the lowest bar that she
might pass.   Suddenly the bundle she clasped
gave a dexterous twist; a small head, with yellow
downy hair, was thrust forth; a pair of fawn-like
eyes fixed an inquiring stare upon him; the pink
face distended with a grin, to which the two small
teeth in the red mouth, otherwise empty, lent a
singularly merry expression; and with a manner
that was a challenge to pursuit, the head disappeared
as suddenly as it had appeared, tucked with affected shyness under Evelina's arm.

She left Stephen standing with the bar in his
hand, staring blankly after her, and ran into the
cabin.

Her father had no questions to ask—nor she.

As he caught her in his arms he gave a great

cry of joy that rang through the house, and brought
Timothy from the barn, in astonishment, to the
scene.

"Eveliny's *home!*" he cried out to Tim, who, with
the ox-yoke in his hand, paused in the doorway.
"Kem ter stay! Eveliny's *home!* I knowed she'd
kem back to her old daddy. Eveliny's kem ter
stay fur good."

"They tole me they'd hectored ye plumb out'n
the town an' out'n yer office. They hed the insurance ter tell *me* that word!" she cried, sobbing
on his breast.

"What d'ye reckon I keer fur enny jestice's
cheer when I hev got ye agin ter set alongside o'
me by the fire?" he exclaimed, his cracked old
voice shrill with triumphant gladness.

He pushed her into her rocking-chair in the
chimney-corner, and laughed again with the supreme pleasure of the moment, although she had
leaned her head against the logs of the wall, and
was sobbing aloud with the contending emotions
that tore her heart.

"Didn't ye ever want ter kem afore, Eveliny?"
he demanded. "I hev been a-pinin' fur a glimge
o' ye." He was in his own place now, his hands
trembling as they lay on the arms of his chair, a
pathetic reproach was in his voice. "Though old
folks oughtn't ter expec' too much o' young ones, ez
be all tuk up naterally with tharse'fs," he added,
bravely. He would not let his past lonely griefs mar
the bright present. "Old folks air mos'ly cumberers—mos'ly cumberers o' the yearth, ennyhow."

Her weeping had ceased; she was looking at him

with dismayed surprise in her eyes, still lustrous with unshed tears. "Why, dad I sent ye a hundred messages ef I mought kem. I tole Abs'lom ter tell Joe Boyd—bein' as ye liked Joe—I wanted ter see ye." She leaned forward and looked up at him with frowning intensity. "They never gin ye that word?"

He laughed aloud in sorry scorn. "We can't teach our chil'n nuthin'," he philosophized. "They hev got ter hurt tharse'fs with all the thorns an' the stings o' the yearth. Our sperience with the sharp things an' bitter ones don't do them no sarvice. Naw, leetle darter—naw! Ye mought ez well gin a message o' kindness ter a wolf, an' expec' him ter kerry it ter some lonesome, helpless thing a-wounded by the way-side, ez gin it ter a Kittredge."

"I never will speak ter one o' 'em agin ez long ez I live," she cried, with a fresh gust of tears.

"Waal," exclaimed the old man, reassuringly, and chirping high, "hyar we all be agin, jes' the same ez we war afore. Don't cry, Eveliny; it's jes' the same."

A sudden babbling intruded upon the conversation. The youthful Kittredge, as he sat upon the wide flat stones of the hearth, was as unwelcome here in the Cove as a Quimbey had been in the cabin on the mountain. The great hickory fire called for his unmixed approval, coming in, as he had done, from the gray wet day. He shuffled his bare pink feet—exceedingly elastic and agile members they seemed to be, and he had a remarkable "purchase" upon their use — and brought them smartly down upon their heels as if this were one of the accepted gestures of applause. Then he looked up at the dark frowning faces of his mother's brothers, and gurgled with laughter, showing the fascinating spectacle of his two front teeth. Perhaps it was the only Kittredge eye that they were not willing to meet. They solemnly gazed beyond him and into the fire, ignoring his very existence. He sustained the slight with an admirable cheerfulness, and babbled and sputtered and flounced about with his hands. He grew pinker in the generous firelight, and he looked very fat as he sat in a heap on the floor. He seemed to have threads tightly tied about his bolster-shaped limbs in places where elder people prefer joints — in his ankles and wrists and elbows—for his arms were bare, and although his frock of pink calico hung decorously high on one shoulder, it drooped quite off from the other, showing a sturdy chest.

His mother took slight notice of him; she was beginning to look about the room with a certain critical disfavor at the different arrangement of the household furniture adopted by her father's deaf and widowed old sister who presided here now, and who, it chanced, had been called away by the illness of a relative. Evelina got up presently, and shifted the position of the spinning-wheels, placing the flax-wheel where the large wheel had been. She then pushed out the table from the corner. "What ailed her ter sot it hyar?" she grumbled, in a disaffected undertone, and shoved it to the centre of the floor, where it had always stood during her own sway. She cast a discerning glance up among the strings of herbs and peppers hanging from above, and ex-

7

amined the shelves where the simple stores for table use were arranged in earthen-ware bowls or gourds —all with an air of vague dissatisfaction. She presently stepped into the shed-room, and there looked over the piles of quilts. They were in order, certainly, but placed in a different method from her own; another woman's hand had been at work, and she was jealous of its very touch among these familiar old things to which she seemed positively akin. "I wonder how I made out ter bide so long on the mounting," she said; and with the recollection of the long-haired Absalom there was another gush of tears and sobs, which she stifled as she could in one of the old quilts that held many of her own stitches and was soothing to touch.

The infantile Kittredge, who was evidently not born to blush unseen, seemed to realize that he had failed to attract the attention of the three absorbed Quimbeys who sat about the fire. He blithely addressed himself to another effort. He suddenly whisked himself over on all-fours, and with a certain ursine aspect went nimbly across the hearth, still holding up his downy yellow head, his pink face agrin, and alluringly displaying his two facetious teeth. He caught the rung of Tim's chair, and lifted himself tremulously to an upright posture. And then it became evident that he was about to give an exhibition of the thrilling feat of walking around a chair. With a truly Kittredge perversity he had selected the one that had the savage Timothy seated in it. For an instant the dark-browed face scowled down into his unaffrighted eyes: it seemed as if Tim might kick him into the fire. The next moment he had set out to circumnavigate, as it were. What a prodigious force he expended upon it! How he gurgled and grinned and twisted his head to observe the effect upon the men, all sedulously gazing into the fire! how he bounced, and anon how he sank with sudden genuflections! how limber his feet seemed, and what free agents! Surely he never intended to put them down at that extravagant angle. More than once one foot was placed on top of the other—an attitude that impeded locomotion and resulted in his sitting down in an involuntary manner and with some emphasis. With an appalling temerity he clutched Tim's great miry boots to help him up and on his way round. Occasionally he swayed to and fro, with his teeth on exhibition, laughing and babbling and shrilly exclaiming, inarticulately bragging of his agile prowess, as if he were able to defy all the Quimbeys, who would not notice him. And when it was all over he went in his wriggling ursine gait back to the hearth-stone, and there he was sitting, demurely enough, and as if he had never moved, when his mother returned and found him.

There was no indication that he had attracted a moment's attention. She looked gravely down at him; then took her chair. A pair of blue yarn socks was in her hand. "I never see sech darnin' ez Aunt Sairy Ann do fur ye, dad; I hev jes tuk my shears an' cut this heel smang out, an' I be goin' ter do it over."

She slipped a tiny gourd into the heel, and began to draw the slow threads to and fro across it.

The blaze, red and yellow, and with elusive purple gleams, leaped up the chimney.   The sap was
still in the wood ; it sang a summer-tide song.   But
an autumn wind was blowing shrilly down the
chimney; one could hear the sibilant rush of the
dead leaves on the blast.   The window and the
door shook, and were still, and once more rattled
as if a hand were on the latch.

Suddenly—" Ever weigh him ?" her father asked.

She sat upright with a nervous start.   It was a
moment before she understood that it was of the
Kittredge scion he spoke.

With his high cracked laugh the old man leaned
over, his outspread hand hovering about the plump
baby, uncertain where, in so much soft fatness, it
might be practicable to clutch him.   There were
some large horn buttons on the back of his frock, a
half-dozen of which, gathered together, afforded a
grasp.   He lifted the child by them, laughing in
undisguised pleasure to feel the substantial strain
upon the garment.

" Toler'ble survigrus," he declared, with his high
chirp.

His daughter suddenly sprang up with a pallid
face and a pointing hand.

" The winder !" she huskily cried—" suthin's at
the winder !"

But when they looked they saw only the dark
square of tiny panes, with the fireside scene genially
reflected on it.   And then she fell to declaring that
she had been dreaming, and besought them not to
take down their guns nor to search, and would not
be still until they had all seemed to concede the

point ; it was she who fastened the doors and shutters, and she did not lie down to rest till they were
all asleep and hours had passed. None of them
doubted that it was Absalom's face that she had
seen at the window, where the light had once lured
him before, and she knew that she had dreamed no
dream like this.

It soon became evident that whenever Joe Boyd
was intrusted with a message he would find means
to deliver it. For upon him presently devolved
the difficult duties of ambassador. The first time
that his honest square face appeared at the rail
fence, and the sound of his voice roused Evelina
as she stood feeding the poultry close by, she returned his question with a counter-question hard
to answer.

"I hev been up the mounting," he said, smiling,
as he hooked his arms over the rail fence. "Abs'-
lom he say he wanter know when ye'll git yer visit
out an' kem home."

She leaned her elbow against the ash-hopper, bal-
ancing the wooden bowl of corn-meal batter on its
edge and trembling a little ; the geese and chickens
and turkeys crowded, a noisy rout, about her feet.

"Joe," she said, irrelevantly, "ye air one o' the
few men on this yearth ez ain't a liar."

He stared at her gravely for a moment, then
burst into a forced laugh. "Ho! ho! I tell a
bushel o' 'em a day, Eveliny!" He wagged his
head in an anxious affectation of mirth.

"Why'n't ye gin dad them messages ez Abs'lom
gin ye from me?"

Joe received this in blank amaze ; then, with sudden comprehension, his lower jaw dropped. He looked at her with a plea for pity in his eyes. And yet his ready tact strove to reassert itself.

" I mus' hev furgot 'em," he faltered.

"Did Abs'lom ever gin 'em ter ye ?" she persisted.

" *Ef he did*, I mus' hev furgot 'em," he repeated, crestfallen and hopeless.

She laughed and turned jauntily away, once more throwing the corn-meal batter to the greedily jostling poultry. "Tell Abs'lom I hev f'und him out," she said. " He can't sot me agin dad no sech way. This be my home, an' hyar I be goin' ter 'bide."

And so she left the good Joe Boyd hooked on by the elbows to the fence.

The Quimbeys, who had heard this conversation from within, derived from it no small elation. " She hev gin 'em the go - by fur good," Timothy said, confidently, to his father, who laughed in triumph, and pulled calmly at his pipe, and looked ten years younger.

But Steve was surlily anxious. " I'd place heap mo' dependence in Eveliny ef she didn't hev this hyar way o' cryin' all the time. She 'lows she's glad she kem—*so glad* she hev lef' Abs'lom fur good an' all—an' then she busts out a-cryin' agin. I ain't able ter argufy on sech."

"Shucks ! wimmen air always a-cryin', an' they don't mean *nuthin'* by it," exclaimed the old man, in the plenitude of his wisdom. "It air jes' one o' thar most contrarious ways. I hev seen 'em set down an' cry fur joy an' pleasure."

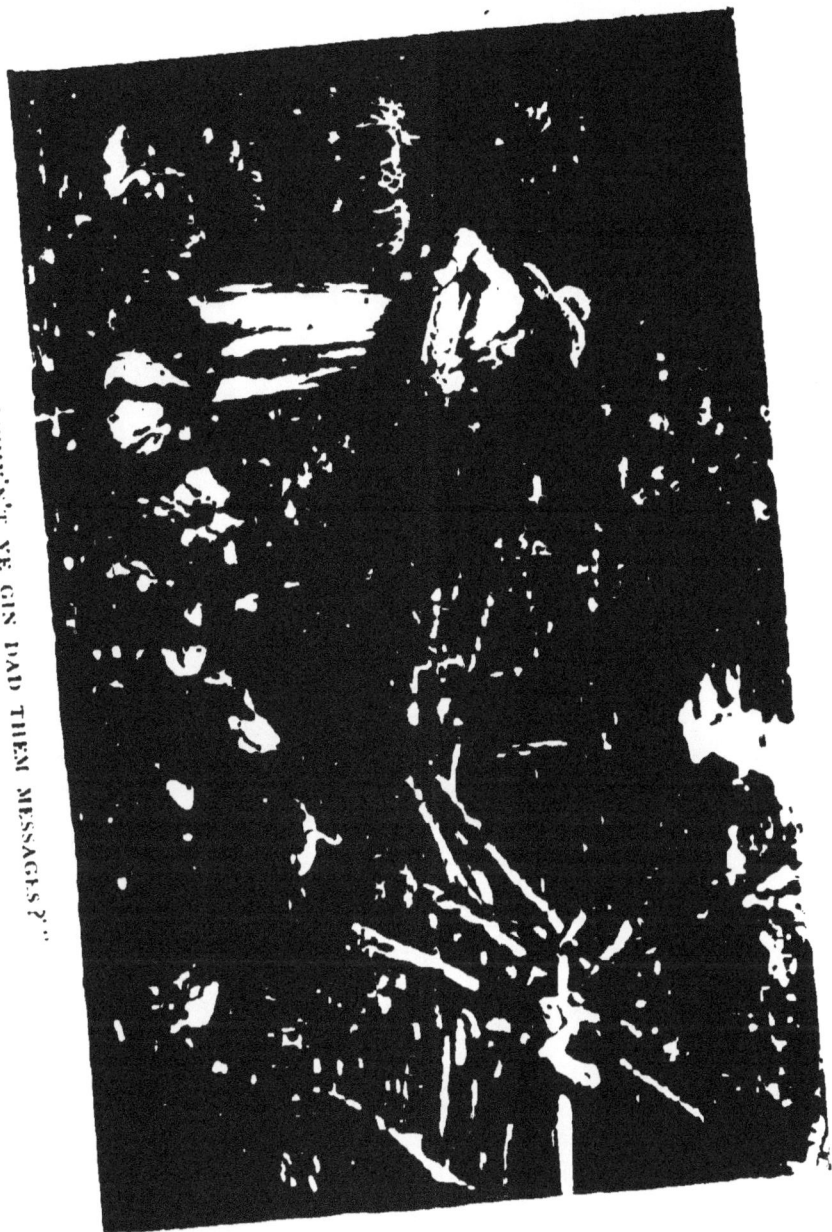

"...WHY'N'T YE GIS DAD THEM MESSAGES?".

But Steve was doubtful. "It be a powerful low-sperited gift fur them ez hev ter 'bide along of 'em. Eveliny never useter be tearful in nowise. Now she cries a heap mo' 'n that thar shoat"—his lips curled in contempt as he glanced toward the door, through which was visible a small rotund figure in pink calico, seated upon the lowest log of the wood-pile—"ez she fotched down hyar with her. *He* never hev hed a reg'lar blate but two or three times sence he hev been hyar, an' them war when that thar old tur-rkey gobbler teetered up ter him an' tuk his corn-dodger that he war a-eatin' on plumb out'n his hand. *He* hed suthin' to holler fur —hed los' his breakfus."

"Don't he 'pear ter you-uns to be powerful peegeon-toed?" asked Tim, anxiously, turning to his father.

"The gawbbler?" faltered the amazed old man.

"Naw; him, *him—Kittredge,*" said Tim, jerking his big thumb in the direction of the small boy.

"Law-dy Gawd A'mighty! *naw! naw!*" The grandfather indignantly repudiated the imputation of the infirmity. One would have imagined that he would deem it meet that a Kittredge should be pigeon-toed. "It's jes the way *all* babies hev got a-walkin'; he ain't right handy yit with his feet— jes a-beginnin' ter walk, an' sech. Peegeon-toed! I say it, ye fool!" He cast a glance of contempt on his eldest-born, and arrogantly puffed his pipe.

Again Joe Boyd came, and yet again. He brought messages contrite and promissory from Absalom; he brought commands stern and insistent. He came into the house at last, and sat and talked at the fireside in the presence of the men of the family, who bore themselves in a manner calculated to impress the Kittredge emissary with their triumph and contempt for his mission, although they studiously kept silence, leaving it to Evelina to answer.

At last the old man, leaning forward, tapped Joe on the knee. "See hyar, Joe. Ye hev always been a good frien' o' mine. This hyar man he stole my darter from me, an' whenst she wanted ter be frien's, an' not let her old dad die unforgivin', he wouldn't let her send the word ter me. An' then he sot himself ter spite an' hector me, an' fairly run· me out'n the town, an' harried me out'n my office; an' when she f'und out — she wouldn't take my word fur it — the deceivin' natur' o' the Kittredge tribe, she hed hed enough o' 'em. I hev let ye argufy 'bout'n it; ye hev hed yer fill of words. An' now I be tired out. Ye ain't 'lowin' she'll ever go back ter her husband, air ye?"

Joe dolorously shook his head.

"Waal, ef ever ye kem hyar talkin' 'bout'n it agin, I'll be 'bleeged ter take down my rifle ter ye."

Joe gazed, unmoved, into the fire.

"An' that would be mighty hard on me, Joe, 'kase ye be so pop'lar 'mongst all, I dunno *what* the kentry-side would do ter me . ef I war ter put a bullet inter ye. Ye air a young man, Joe. Ye oughter spare a old man sech a danger ez that."

And so it happened that Joe Boyd's offices as mediator ceased.

A week went by in silence and without result.

Evelina's tears seemed to keep count of the minutes. The brothers indignantly noted it, and even the old man was roused from the placid securities of his theories concerning lachrymose womankind, and remonstrated sometimes, and sometimes grew angry and exhorted her to go back. What did it matter to her how her father was treated? He was a cumberer of the ground, and many people besides her husband had thought he had no right to sit in a justice's chair. And then she would burst into tears once more, and declare again that she would never go back.

The only thoroughly cheerful soul about the place was the intruding Kittredge. He sat continuously—for the weather was fine—on the lowest log of the wood-pile, and swung his bare pink feet among the chips and bark, and seemed to have given up all ambition to walk. Occasionally red and yellow leaves whisked past his astonished eyes, although these were few now, for November was on the wane. He babbled to the chickens, who pecked about him with as much indifference as if he were made of wood. His two teeth came glittering out whenever the rooster crowed, and his gleeful laugh —he rejoiced so in this handsomely endowed bird —could be heard to the barn. The dogs seemed never to have known that he was a Kittredge, and wagged their tails at the very sound of his voice, and seized surreptitious opportunities to lick his face. Of all his underfoot world only the gobbler awed him into gravity and silence; he would gaze in dismay as the marauding fowl irresolutely approached from around the wood-pile, with long

neck out-stretched and undulating gait, applying
first one eye and then the other to the pink hands,
for the gobbler seemed to consider them a perpetual repository of corn-dodgers, which indeed
they were. Then the head and the wabbling red
wattles would dart forth with a sudden peck, and
the shriek that ensued proved that nothing could
be much amiss with the Kittredge lungs.

  One fine day he sat thus in the red November
sunset. The sky, seen through the interlacing
black boughs above his head, was all amber and
crimson, save for a wide space of pure and pallid
green, against which the purplish-garnet wintry
mountains darkly gloomed. Beyond the rail fence
the avenues of the bare woods were carpeted with
the sere yellowish leaves that gave back the sunlight with a responsive illuminating effect, and thus
the sylvan vistas glowed. The long slanting beams
elongated his squatty little shadow till it was hardly a caricature. He heard the cow lowing as she
came to be milked, fording the river where the
clouds were so splendidly reflected. The chickens
were going to roost. The odor of the wood, the
newly-hewn chips, imparted a fresh and fragrant
aroma to the air. He had found among them a
sweet-gum ball and a pine cone, and was applying
them to the invariable test of taste. Suddenly he
dropped them with a nervous start, his lips trembled, his lower jaw fell, he was aware of a stealthy
approach. Something was creeping behind the
wood-pile. He hardly had time to bethink himself
of his enemy the gobbler when he was clutched
under the arm, swung through the air with a swift-

ness that caused the scream to evaporate in his
throat, and the next moment he looked quakingly
up into his father's face with unrecognizing eyes;
for he had forgotten Absalom in these few weeks.
He squirmed and wriggled as he was held on the
pommel of the saddle, winking and catching his
breath and spluttering, as preliminary proceedings
to an outcry. There was a sudden sound of heavily shod feet running across the puncheon floor
within, a wild, incoherent exclamation smote the
air, an interval of significant silence ensued.

"Get up!" cried Absalom, not waiting for Tim's
rifle, but spurring the young horse, and putting
him at the fence. The animal rose with the elas-
ticity and lightness of an uprearing ocean wave.
The baby once more twisted his soft neck, and
looked anxiously into the rider's face. This was
not the gobbler. The gobbler did not ride horse-
back. Then the affinity of the male infant for the
noble equine animal suddenly overbore all else.
In elation he smote with his soft pink hand the
glossy arched neck before him. "Dul-lup!" he
arrogantly echoed Absalom's words. And thus
father and son at a single bound disappeared into
woods, and so out of sight.

The savage Tim was leaning upon his rifle in
the doorway, his eyes dilated, his breath short, his
whole frame trembling with excitement, as the
other men, alarmed by Evelina's screams, rushed
down from the barn.

"What ails ye, Tim? Why'n't ye fire?" demanded his father.

Tim turned an agitated, baffled look upon him.
"I—I mought hev hit the baby," he faltered.

"Hain't ye got no aim, ye durned sinner?" asked
Stephen, furiously.

"Bullet mought hev gone through him and struck
inter the baby," expostulated Tim.

"An' then agin it moughtn't!" cried Stephen.
"Lawd, ef *I* hed hed the chance!"

"Ye wouldn't hev done no differ," declared Tim.

"Hyar!" Steve caught his brother's gun and
presented it to Tim's lips. "Suck the bar'l. It's
'bout all ye air good fur."

The horses had been turned out. By the time
they were caught and saddled pursuit was evidently
hopeless. The men strode in one by one, dashing
the saddles and bridles on the floor, and finding in
angry expletives a vent for their grief. And indeed
it might have seemed that the Quimbeys must have
long sought a choice Kittredge infant for adop-
tion, so far did their bewailings discount Rachel's
mourning.

"Don't cry, Eveliny," they said, ever and anon.
"We-uns 'll git him back fur ye."

But she had not shed a tear. She sat speech-
less, motionless, as if turned to stone.

"Laws-a-massy, child, ef ye would jes hev
b'lieved *me* 'bout'n them Kittredges — Abs'lom in
partic'lar — ye'd be happy an' free now," said the
old man, his imagination somewhat extending his
experience, for he had had no knowledge of his
son-in-law until their relationship began.

The evening wore drearily on. Now and then
the men roused themselves, and with lowering

faces discussed the opportunities of reprisal, and the best means of rescuing the child. And whether they schemed to burn the Kittredge cabin, or to arm themselves, burst in upon their enemies, shooting and killing all who resisted, Evelina said nothing, but stared into the fire with unnaturally dilated eyes, her white lined face all drawn and somehow unrecognizable.

"Never mind," her father said at intervals, taking her cold hand, " we-uns 'll git him back, Eveliny. The Lord hed a mother wunst, an' I'll be bound He keeps a special pity for a woman an' her child."

"Oh, great gosh ! who'd hev dreamt we'd hev missed him so !" cried Tim, shifting his position, and slipping his left arm over the back of his chair. "Jes ter think o' the leetle size o' him, an' the great big gap he hev lef' roun' this hyar ha'th-stone !"

" An' yit he jes sot underfoot, 'mongst the cat an' the dogs, jes ez humble !" said Stephen.

" I'd git him back even ef he warn't no kin ter me, Eveliny," declared Tim, and he spoke advisedly, remembering that the youth was a Kittredge.

Still Evelina said not a word. All that night she silently walked the puncheon floor, while the rest of the household slept. The dogs, in vague disturbance, because of the unprecedented vigil and stir in the midnight, wheezed uneasily from time to time, and crept restlessly about under the cabin, now and again thumping their backs or heads against the floor; but at last they betook themselves to slumber. The hickory logs broke in

twain as they burned, and fell on either side, and
presently there was only the dull red glow of the
embers on her pale face, and the room was full of
brown shadows, motionless, now that the flames
flared no more.   Once when the red glow, growing
ever dimmer, seemed almost submerged beneath
the gray ashes, she paused and stirred the coals.
The renewed glimmer showed a fixed expression
in her eyes, becoming momently more resolute.
At intervals she knelt at the window and placed
her hands about her face to shut out the light
from the hearth, and looked out upon the night.
How the chill stars loitered !   How the dawn
delayed !   The great mountain gloomed darkling
above the Cove.   The waning moon, all melancholy
and mystic, swung in the purple sky.   The bare,
stark boughs of the trees gave out here and there
a glimmer of hoar-frost.   There was no wind ;
when she heard the dry leaves whisk she caught a
sudden glimpse of a fox that, with his crafty shadow
pursuing him, leaped upon the wood-pile, nimbly
ran along its length, and so, noiselessly, away —
while the dogs snored beneath the house.   A cock
crew from the chicken-roost ; the mountain echoed
the resonant strain.   She saw a mist come stealing softly along a precipitous gorge ; the gauzy
web hung shimmering in the moon ; presently
the trees were invisible ; anon they showed rigid
among the soft enmeshment of the vapor, and again
were lost to view.

She rose ; there was a new energy in her step ;
she walked quickly across the floor and unbarred
the door.

The little cabin on the mountain was lost among
the clouds. It was not yet day, but the old woman,
with that proclivity to early rising characteristic of
advancing years, was already astir. It was in the
principal room of the cabin that she slept, and it
contained another bed, in which, placed crosswise,
were five billet-shaped objects under the quilts,
which when awake identified themselves as Peter
Kittredge's children. She had dressed and un-
covered the embers, and put on a few of the chips
which had been spread out on the hearth to dry,
and had sat down in the chimney corner. A timid
blaze began to steal up, and again was quenched,
and only the smoke ascended in its form; then the
light flickered out once more, casting a gigantic
shadow of her sun-bonnet—for she had donned it
thus early—half upon the brown and yellow daubed
wall, and half upon the dark ceiling, making a specious stir amidst the peltry and strings of pop-corn
hanging motionless thence.

She sighed heavily once or twice, and with an
aged manner, and leaned her elbows on her knees
and gazed contemplatively at the fire. All at once
the ashes were whisked about the hearth as in a
sudden draught, and then were still. In momentary
surprise she pushed her chair back, hesitated, then
replaced it, and calmly settled again her elbows on
her knees. Suddenly once more a whisking of the
ashes; a cold shiver ran through her, and she
turned to see a hand fumbling at the batten shutter
close by. She stared for a moment as if paralyzed;
her spectacles fell to the floor from her nerveless
hand, shattering the lenses on the hearth. She

rose trembling to her feet, and her lips parted as if
to cry out. They emitted no sound, and she turned
with a terrified fascination and looked back. The
shutter had opened; there was no glass; the small
square of the window showed the nebulous gray
mist without, and defined upon it was Evelina's
head, her dark hair streaming over the red shawl
held about it, her fair oval face pallid and pensive,
and with a great wistfulness upon it; her lustrous
dark eyes glittered.

"Mother," her red lips quivered out.

The old crone recognized no treachery in her
heart. She laid a warning finger upon her lips.
All the men were asleep.

Evelina stretched out her yearning arms. "Gin
him ter me!"

"Naw, naw, Eveliny," huskily whispered Absalom's mother. "Ye oughter kem hyar an' 'bide
with yer husband—ye know ye ought."

Evelina still held out her insistent arms. "Gin
him ter me!" she pleaded.

The old woman shook her head sternly. "Ye
kem in, an' 'bide whar ye b'long."

Evelina took a step nearer the window. She
laid her hand on the sill. "Spos'n 'twar Abs'lom
whenst he war a baby," she said, her eyes softly
brightening, "an' another woman hed him an' kep'
him, 'kase ye an' his dad fell out—would ye hev
'lowed she war right ter treat ye like ye treat me—
whenst Abs'lom war a baby?"

Once more she held out her arms.

There was a step in the inner shed-room; then
silence.

" Ye hain't got no excuse," the soft voice urged;
" ye know jes how I feel, how ye'd hev felt, whenst
Abs'lom war a baby."

The shawl had fallen back from her tender face;
her eyes glowed, her cheek was softly flushed. A
sudden terror thrilled through her as she again
heard the heavy step approaching in the shed-room.
" Whenst Abs'lom war a baby," she reiterated, her
whole pleading heart in the tones.

A sudden radiance seemed to illumine the sad,
dun-colored folds of the encompassing cloud; her
face shone with a transfiguring happiness, for the
hustling old crone had handed out to her a warm,
somnolent bundle, and the shutter closed upon the
mists with a bang.

" The wind's riz powerful suddint," Peter said,
noticing the noise as he came stumbling in, rubbing
his eyes. He went and fastened the shutter, while
his mother tremulously mended the fire.

The absence of the baby was not noticed for
some time, and when the father's hasty and angry
questions elicited the reluctant facts, the outcry for
his loss was hardly less bitter among the Kittredges
than among the Quimbeys. The fugitives were
shielded from capture by the enveloping mist, and
when Absalom returned from the search he could
do naught but indignantly upbraid his mother.

She was terrified by her own deed, and cowered
under Absalom's wrath. It was in a moral collapse,
she felt, that she could have done this thing. She
flung her apron over her head, and sat still and
silent—a monumental figure—among them. Once,
roused by Absalom's reproaches, she made some

8

effort to defend and exculpate herself, speaking
from behind the enveloping apron.

"I ain't born no Kittredge nohow," she irrelevantly asseverated, "an' I never war.   An' when
Eveliny axed me how I'd hev liked ter hev another
'oman take Abs'lom whenst he war a baby, I couldn't
hold out no longer."

"Shucks!" cried Absalom, unfilially; "ye'd a heap
better be a-studyin' 'bout'n my good now 'n whenst
I war a baby—a-givin' away *my* child ter them
Quimbeys ; a-h'istin' him out'n the winder!"

She was glad to retort that he was "impident,"
and to take refuge in an aggrieved silence, as many
another mother has done when outmatched by logic.

After this there was more cheerfulness in her hidden face than might have been argued from her
port of important sorrow.   "Bes' ter hev no jawin',
though," she said to herself, as she sat thus inscrut-
ably veiled.   And deep in her repentant heart she
was contradictorily glad that Evelina and the baby
were safe together down in the Cove.

Old Joel Quimbey, putting on his spectacles, with
a look of keenest curiosity, to read a paper which
the deputy-sheriff of the county presented when he
drew rein by the wood-pile one afternoon some
three weeks later, had some difficulty in identifying
a certain Elnathan Daniel Kittredge specified therein. He took off his spectacles, rubbed them smartly, and put them on again. The writing was unchanged. Surely it must mean the baby. That
was the only Kittredge whose body they could be
summoned to produce on the 24th of December

before the judge of the circuit court, now in session. He turned the paper about and looked at it, his natural interest as a man augmented by his recognition as an ex-magistrate of its high important legal character.

"Eveliny," he quavered, at once flattered and furious, "dad-burned ef Abs'lom hain't gone an' got out a *habeas corpus* fur the baby!"

The phrase had a sound so deadly that there was much ado to satisfactorily explain the writ and its functions to Evelina, who had felt at ease again since the baby was at home, and so effectually guarded that to kidnap him was necessarily to murder two or three of the vigilant and stalwart Quimbey men. So much joy did it afford the old man to air his learning and consult his code—a relic of his justiceship—that he belittled the danger of losing the said Elnathan Daniel Kittredge in the interest with which he looked forward to the day for him to be produced before the court.

There was a gathering of the clans on that day. Quimbeys and Kittredges who had not visited the town for twenty years were jogging thither betimes that morning on the red clay roads, all unimpeded by the deep mud which, frozen into stiff ruts and ridges here and there, made the way hazardous to the running-gear. The lagging winter had come, and the ground was half covered with a light fall of snow.

The windows of the court-house were white with frost; the weighted doors clanged continuously. An old codger, slowly ascending the steps, and pushing into the semi-obscurity of the hall, paused

as the door slammed behind him, stared at the
sheriff in surprise, then fixed him with a bantering
leer.   The light that slanted through the open
court-room door fell upon the official's burly figure,
his long red beard, his big broad-brimmed hat
pushed back from his laughing red face, consciously ludicrous and abashed just now.

"Hev ye made a find?" demanded the new-
comer.   .

For in the strong arms of the law sat, bolt-
upright, Elnathan Daniel Kittredge, his yellow
head actively turning about, his face decorated
with a grin, and on most congenial terms with
the sheriff.

"They're lawin' 'bout'n him in thar"—the sheriff
jerked his thumb toward the door.   "*Habeas corpus*
perceedin's.   Dunno ez I ever see a friskier leetle
cuss.   Durned ef I 'ain't got a good mind ter run
off with him myself."

The said Elnathan Daniel Kittredge once more
squirmed round and settled himself comfortably in
the hollow of the sheriff's elbow, who marvelled to
find himself so deft in holding him, for it was twenty years since his son—a gawky youth who now
affected the company at the saloon, and was none
too filial—was the age and about the build of this
infant Kittredge.

"They hed a reg'lar scrimmage hyar in the hall
—them fool men—Quimbey an' Kittredge.   Old
man Quimbey said suthin' ter Abs'lom Kittredge
—I dunno what all.   Abs'lom never jawed back
none.   He jes made a dart an' snatched this hyar
leetle critter out'n his mother's arms, stiddier wait-

in' fur the law, what he summonsed himself. Blest
ef I didn't hev ter hold my revolver ter his head,
an' then crack him over the knuckles, ter make him
let go the child.   I didn't want ter arrest him—
mighty clever boy, Abs'lom Kittredge!   I promised
that young woman I'd keep holt o' the child till the
law gins its say-so.   I feel sorry fur her; she's been
through a heap."

"Waal, ye look mighty pritty, totin' him around
hyar," his friend encouraged him with a grin. "I'll
say that fur ye—ye look mighty pritty."

And in fact the merriment in the hall at the sheriff's expense began to grow so exhilarating as to
make him feel that the proceedings within were too
interesting to lose. His broad red face with its big
red beard reappeared in the doorway—slightly embarrassed because of the sprightly manners of his
charge, who challenged to mirth every eye that
glanced at him by his toothful grin and his gurgles
and bounces; he was evidently enjoying the excitement and his conspicuous position. He manfully
gnawed at his corn-dodger from time to time, and
from the manner in which he fraternized with his
new acquaintance, the sheriff, he seemed old enough
to dispense with maternal care, and, but for his in-
complete methods of locomotion, able to knock about
town with the boys. The Quimbeys took note of
his mature demeanor with sinking hearts; they
looked anxiously at the judge, wondering if he had
ever before seen such precocity—anything so young
to be so old: "He 'ain't never afore 'peared so
survigrus — so *durned* survigrus ez he do ter-day,"
they whispered to each other.

"Yes, sir," his father was saying, on examination, "year old. Eats anything he kin git—cab-
bage an' fat meat an' anything. *Could* walk if he
wanted ter. But he 'ain't been raised right"—he
glanced at his wife to observe the effect of this
statement. He felt a pang as he noted her pensive,
downcast face, all tremulous and agitated, over-
whelmed as she was by the crowd and the infinite
moment of the decision. But Absalom, too, had his
griefs, and they expressed themselves perversely.

"He hev been pompered an' fattened by bein' let
ter eat an' sleep so much, till he be so heavy ter
his self he don't wanter take the trouble ter git
about. He *could* walk ennywhar. He's plumb
survigrus."

And as if in confirmation, the youthful Kittredge
lifted his voice to display his lung power. He hilariously babbled, and suddenly roared out a stento-
rian whoop, elicited by nothing in particular, then
caught the sheriff's beard, and buried in it his conscious pink face.

The judge looked gravely up over his spectacles.
He had a bronzed complexion, a serious, pondering
expression, a bald head, and a gray beard. He wore
a black broadcloth suit, somewhat old-fashioned in
cut, and his black velvet waist-coat had suffered an
eruption of tiny red satin spots. He had great respect for judicial decorums, and no Kittredge, however youthful, or survigrus, or exalted in importance
by *habeas corpus* proceedings, could "holler" unmolested where he presided.

"Mr. Sheriff," he said, solemnly, "remove that
child from the presence of the court."

And the said Elnathan Daniel Kittredge went
out gleefully kicking in the arms of the law.

The hundred or so grinning faces in the court-
room relapsed quickly into gravity and excited interest. The rows of jeans-clad countrymen seated
upon the long benches on either side of the bar
leaned forward with intent attitudes. For this was
a rich feast of local gossip, such as had not been so
bountifully spread within their recollection. All
the ancient Quimbey and Kittredge feuds contrived
to be detailed anew in offering to the judge reasons
why father or mother was the more fit custodian of
the child in litigation.

As Absalom sat listening to all this, his eyes
were suddenly arrested by his wife's face — half
draped it was, half shadowed by her sun-bonnet,
its fine and delicate profile distinctly outlined
against the crystalline and frosted pane of the window near which she sat. The snow without threw
a white reflection upon it; its rich coloring in contrast was the more intense; it was very pensive,
with the heavy lids drooping over the lustrous eyes,
and with a pathetic appeal in its expression.

And suddenly his thoughts wandered far afield.
He wondered that it had come to this; that she
could have misunderstood him so; that he had
thought her hard and perverse and unforgiving.
His heart was all at once melting within him;
somehow he was reminded how slight a thing she
was, and how strong was the power that nerved her
slender hand to drag his heavy weight, in his dead
and helpless unconsciousness, down to the bars
and into the safety of the sheltering laurel that

night, when he lay wounded and bleeding under
the lighted window of the cabin in the Cove. A
deep tenderness, an irresistible yearning had come
upon him; he was about to rise, he was about to
speak he knew not what, when suddenly her face
was irradiated as one who sees a blessed vision ; a
happy light sprang into her eyes; her lips curved
with a smile; the quick tears dropped one by one
on her hands, nervously clasping and unclasping
each other. He was bewildered for a moment.
Then he heard Peter gruffly growling a half-whis-
pered curse, and the voice of the judge, in the exercise of his discretion, methodically droning out his
reasons for leaving so young a child in the custody
of its mother, disregarding the paramount rights of
the father. The judge concluded by dispassionately recommending the young couple to betake themselves home, and to try to live in peace together, or,
at any rate, like sane people. Then he thrust his
spectacles up on his forehead, drew a long sigh of
dismissal, and said, with a freshened look of interest, "Mr. Clerk, call the next case."

The Quimbey and Kittredge factions poured into
the hall; what cared they for the disputed claims
of Jenkins *versus* Jones? The lovers of sensation
cherished a hope that there might be a lawless effort to rescue the infant Kittredge from the custody to which he had been committed by the court.
The Quimbeys watchfully kept about him in a close
squad, his pink sun-bonnet, in which his head was
eclipsed, visible among their brawny jeans shoulders, as his mother carried him in her arms. The
sheriff looked smilingly after him from the court-

house steps, then inhaled a long breath, and began
to roar out to the icy air the name of a witness
wanted within. Instead of a gate there was a flight
of steps on each side of the fence, surmounted by
a small platform. Evelina suddenly shrank back
as she stood on the platform, for beside the fence
Absalom was waiting. Timothy hastily vaulted
over the fence, drew his "shooting-iron" from his
boot-leg, and cocked it with a metallic click, sharp
and peremptory in the keen wintry air. For a moment Absalom said not a word. He looked up at
Evelina with as much reproach as bitterness in his
dark eyes. They were bright with the anger that
fired his blood; it was hot in his bronzed cheek; it
quivered in his hands. The dry and cold atmosphere amplified the graces of his long curling yellow hair that she and his mother loved. His hat
was pushed back from his face. He had not
spoken to her since the day of his ill-starred confidence, but he would not be denied now.

"Ye'll repent it," he said, threateningly. "I'll
take special pains fur that."

She bestowed on him one defiant glance, and
laughed—a bitter little laugh. "Ye air ekal ter it;
ye have a special gift fur makin' folks repent they
ever seen ye."

"The jedge jes gin him ter ye 'kase ye made him
out sech a fibble little pusson," he sneered. "But
it's jes fur a time."

She held the baby closer. He busied himself in
taking off his sun-bonnet and putting it on hind part
before, gurgling with smothered laughter to find
himself thus queerly masked, and he made futile

efforts to play "peep-eye" with anybody jovially
disposed in the crowd.   But they were all gravely
absorbed in the conjugal quarrel at which they
were privileged to assist.

"It's jes fur a time," he reiterated.

"Wait an' see !" she retorted, triumphantly.

"I won't wait," he declared, goaded; "I'll take
him yit; an' when I do I'll clar out'n the State o'
Tennessee—see ef I don't !"

She turned white and trembled.   "Ye dassent,"
she cried out shrilly.   "Ye'll be 'feared o' the law."

"Wait an' see !" He mockingly echoed her words,
and turned in his old confident manner, and strode
out of the crowd.

Faint and trembling, she crept into the old can-
vas-covered wagon, and as it jogged along down
the road stiff with its frozen ruts and ever nearing
the mountains, she clasped the cheerful Kittredge
with a yearning sense of loss, and declared that
the judge had made him no safer than before.   It
was in vain that her father, speaking from the
legal lore of the code, detailed the contempt of
court that the Kittredges would commit should
they undertake to interfere with the judicial decision—it might be even considered kidnapping.

"But what good would that do me—an' the baby
whisked plumb out'n the State?   Ef Abs'lom ain't
'feared o' Tim's rifle, what's he goin' ter keer fur
the pore jedge with nare weepon but his leetle contempt o' court — ter jail Abs'lom, ef he kin make
out ter ketch him !"

She leaned against the swaying hoop of the cover of the wagon and burst into tears.   "Oh, none

o' ye 'll do nuthin' fur me !" she exclaimed, in frantic reproach. " Nuthin' !"

" Ye talk like 'twar we-uns ez made up sech foolishness ez *habeas corpus* out'n our own heads," said Timothy. " I 'ain't never looked ter the law fur pertection. Hyar's the pertecter." He touched the trigger of his rifle and glanced reassuringly at his sister as he sat beside her on the plank laid as a seat from side to side of the wagon.

She calmed herself for a moment; then suddenly looked aghast at the rifle, and with some occult and hideous thought, burst anew into tears.

" Waal, sir," exclaimed Stephen, outdone, " what with all this hyar daily weepin' an' nightly mournin', I 'ain't got spunk enough lef' ter stan' up agin the leetlest Kittredge a-goin'. I ain't man enough ter sight a rifle. Kittredges kin kem enny time an' take my hide, horns, an' tallow ef they air minded so ter do."

" I 'lowed I hearn suthin' a-gallopin' down the. road," said Tim, abruptly.

Her tears suddenly ceased. She clutched the baby closer, and turned and lifted the flap of the white curtain at the back of the wagon, and looked out with a wild and terror-stricken eye. The red clay road stretched curveless, a long way visible and vacant. The black bare trees stood shivering in the chilly blast on either side; among them was an occasional clump of funereal cedars. Away off the brown wooded hills rose; snow lay in thin crust-like patches here and there, and again the earth wore the pallid gray of the crab-grass or the ochreous red of the gully-washed clay.

"I don't see nuthin'," she said, in the bated voice of affrighted suspense.

While she still looked out flakes suddenly began to fly, hardly falling at first, but poised tentatively, fluctuating athwart the scene, presently thickening, quickening, obscuring it all, isolating the woods with an added sense of solitude since the sight of the world and the sound of it were so speedily annulled. Even the creak of the wagon-wheels was muffled. Through the semicircular aperture in the front of the wagon-cover the horns of the oxen were dimly seen amidst the serried flakes; the snow whitened the backs of the beasts and added its burden to their yoke. Once as they jogged on she fancied again that she heard hoof-beats—this time a long way ahead, thundering over a little bridge high above a swirling torrent, that reverberated with a hollow tone to the faintest footfall. "Jes somebody ez hev passed we-uns, takin' the short-cut by the bridle-path," she ruminated. No pursuer, evidently.

Everything was deeply submerged in the snow before they reached the dark little cabin nestling in the Cove. Motionless and dreary it was; not even a blue and gauzy wreath curled out of the chimney, for the fire had died on the hearth in their absence. No living creature was to be seen. The fowls were huddled together in the hen-house, and the dogs had accompanied the family to town, trotting beneath the wagon with lolling tongues and smoking breath; when they nimbly climbed the fence their circular footprints were the first traces to mar the level expanse of the door-yard. The

"HE STOLE NOISELESSLY IN THE SOFT SNOW"

bare limbs of the trees were laden; the cedars bore great flower-like tufts amidst the interlacing fibrous foliage. The eaves were heavily thatched; the drifts lay in the fence corners.

Everything was covered except, indeed, one side of the fodder-stack that stood close to the barn. Evelina, going out to milk the cow, gazed at it for a moment in surprise. The snow had slipped down from it, and lay in rolls and piles about the base, intermixed with the sere husks and blades that seemed torn out of the great cone. "Waal, sir, Spot mus' hev been hongry fur true, ter kem a-foragin' this wise. Looks ez ef she hev been fairly a-burrowin'."

She turned and glanced over her shoulder at tracks in the snow — shapeless holes, and filling fast — which she did not doubt were the footprints of the big red cow, standing half in and half out of the wide door, slowly chewing her cud, her breath visibly curling out on the chill air, her great lips opening to emit a muttered low. She moved forward suddenly into the shelter as Evelina started anew toward it, holding the piggin in one hand and clasping the baby in the other arm.

Evelina noted the sound of her brothers' two axes, busy at the wood-pile, their regular cleavage splitting the air with a sharp stroke and bringing a crystalline shivering echo from the icy mountain. She did not see the crouching figure that came cautiously burrowing out from the stack. Absalom rose to his full height, looking keenly about him the while, and stole noiselessly in the soft snow to the stable, and peered in through a crevice in the wall.

Evelina had placed the piggin upon the straw-covered ground, and stood among the horned cattle and the huddling sheep, her soft melancholy face half shaded by the red shawl thrown over her head and shoulders. A tress of her brown hair escaped and curled about her white neck, and hung down over the bosom of her dark-blue homespun dress. Against her shoulder the dun-colored cow rubbed her horned head. The baby was in a pensive mood, and scarcely babbled. The reflection of the snow was on his face, heightening the exquisite purity of the tints of his infantile complexion. His gentle, fawn-like eyes were full of soft and lustrous languors. His long lashes drooped over them now, and again were lifted. His short down of yellow hair glimmered golden against the red shawl over his mother's shoulders.

One of the beasts sank slowly upon the ground — a tired creature doubtless, and night was at hand ; then another, and still another. Their posture reminded Absalom, as he looked, that this was Christmas Eve, and of the old superstition that the cattle of the barns spend the night upon their knees, in memory of the wondrous Presence that once graced their lowly place. The boughs rattled suddenly in the chill blast above his head ; the drifts fell about him. He glanced up mechanically to see in the zenith a star of gracious glister, tremulous and tender, in the rifts of the breaking clouds.

"I wonder ef it air the same star o' Bethlehem?" he said, thinking of the great sidereal torch heralding the Light of the World. He had a vague sense that this star has never set, however the wan-

dering planets may come and go in their wide journeys as the seasons roll. He looked again into the
glooming place, at the mother and her child, remembering that the Lord of heaven and earth had
once lain in a manger, and clung to a humble
earthly mother.

The man shook with a sudden affright. He had
intended to wrest the child from her grasp, and
mount and ride away; he was roused from his
reverie by the thrusting upon him of his opportunity, facilitated a hundredfold. Evelina had evidently forgotten something. She hesitated for a
moment; then put the baby down upon a great pile
of straw among the horned creatures, and, catching
her shawl about her head, ran swiftly to the house.

Absalom moved mechanically into the doorway.
The child, still pensive and silent, and looking tenderly infantile, lay upon the straw. A sudden
pang of pity for her pierced his heart: how her own
would be desolated! His horse, hitched in a clump
of cedars, awaited him ten steps away. It was his
only chance—his last chance. And he had been
hardly entreated. The child's eyes rested, startled
and dilated, upon him; he must be quick.

The next instant he turned suddenly, ran hastily through the snow, crashed among the cedars,
mounted his horse, and galloped away.

It was only a moment that Evelina expected to
be at the house, but the gourd of salt which she
sought was not in its place. She hurried out with
it at last, unprescient of any danger until all at
once she saw the footprints of a man in the snow,
otherwise untrodden, about the fodder-stack. She

still heard the two axes at the wood-pile. Her
father, she knew, was at the house.

A smothered scream escaped her lips. The steps
had evidently gone into the stable, and had come
out thence. Her faltering strength could scarcely
support her to the door. And then she saw lying
in the straw Elnathan Daniel, beginning to babble
and gurgle again, and to grow very pink with joy
over a new toy—a man's glove, a red woollen glove,
accidentally dropped in the straw. She caught it
from his hands, and turned it about curiously. She
had knit it herself—for Absalom !

When she came into the house, beaming with joy,
the baby holding the glove in his hands, the men
listened to her in dumfounded amaze, and with significant side glances at each other.

"He wouldn't take the baby whenst he hed the
chance, 'kase he knowed 'twould hurt me so. An'
he never wanted ter torment me — I reckon he
never *did* mean ter torment me. An' he did 'low
wunst he war sorry he spited dad. Oh ! I hev been
a heap too quick an' spiteful myself. I hev been
so terrible wrong ! Look a-hyar ; he lef' this glove
ter show me he hed been hyar, an' could hev tuk
the baby ef he hed hed the heart ter do it. Oh !
I'm goin' right up the mounting an' tell him how
sorry I be."

"Toler'ble cheap !" grumbled Stephen — "one
old glove. An' he'll git Elnathan Daniel an' ye too.
A smart fox he be."

They could not dissuade her. And after a time
it came to pass that the Quimbey and Kittredge
feuds were healed ; for how could the heart of a

OLD QUIMBEY AND HIS GRANDSON

grandfather withstand a toddling spectacle in pink calico that ran away one day some two years later, in company with an adventurous dog, and came down the mountain to the cabin in the Cove, squeezing through the fence rails after the manner of his underfoot world, proceeding thence to the house, where he made himself very merry and very welcome? And when Tim mounted his horse and rode up the mountain with the youngster on the pommel of the saddle, lest Evelina should be out of her mind with fright because of his absence, how should he and old Mrs. Kittredge differ in their respective opinions of his vigorous growth, and grace of countenance, and peartness of manner? On the strength of this concurrence Tim was induced to "'light an' hitch," and he even sat on the cabin porch and talked over the crops with Absalom, who, the next time he went to town, stopped at the cabin in the Cove to bring word how Elnathan Daniel was " thrivin'." The path that Evelina had worn to the crag in those first homesick days on the mountain rapidly extended itself into the Cove, and widened and grew smooth, as the grandfather went up and the grandson came down.

9

ONE memorable night in Lonesome Cove the ranger of the county entered upon a momentous crisis in his life. What hour it was he could hardly have said, for the primitive household reckoned time by the sun when it shone, by the domestic routine when no better might be. It was late. The old crone in the chimney-corner nodded over her knitting. In the trundle-bed at the farther end of the shadowy room were transverse billows under the quilts, which intimated that the small children were numerous enough for the necessity of sleeping crosswise. He had smoked out many pipes, and at last knocked the cinder from the bowl. The great hickory logs had burned asunder and fallen from the stones that served as andirons. He began to slowly cover the embers with ashes, that the fire might keep till morning.

His wife, a faded woman, grown early old, was bringing the stone jar of yeast to place close by the hearth, that it might not " take a chill " in some sudden change of the night. It was heavy, and she bent in carrying it. Awkward, and perhaps nervous, she brought it sharply against the shovel in his hands.

The clash roused the old crone in the corner.

She recognized the situation instantly, and the features that sleep had relaxed into inexpressiveness
took on a weary apprehension, which they wore like
a habit. The man barely raised his surly black
eyes, but his wife drew back humbly with a mutter
of apology.

The next moment the shovel was almost thrust
out of his grasp. A tiny barefooted girl, in a
straight unbleached cotten night-gown and a quaint
little cotton night-cap, cavalierly pushed him aside,
that she might cover in the hot ashes a burly sweet-
potato, destined to slowly roast by morning. A
long and careful job she made of it, and unconcern-
edly kept him waiting while she pottered back and
forth about the hearth. She loooked up once with
an authoritative eye, and he hastily helped to ad-
just the potato with the end of the shovel. And
then he glanced at her, incongruously enough, as
if waiting for her autocratic nod of approval. She
gravely accorded it, and pattered nimbly across the
puncheon floor to the bed.

"Now," he drawled, in gruff accents, "ef you-uns
hev all had yer fill o' foolin' with this hyar fire, I'll
kiver it, like I hev started out ter do."

At this moment there was a loud trampling upon
the porch without. The batten door shook violently.
The ranger sprang up. As he frowned the hair on
his scalp, drawn forward, seemed to rise like bristles.

"Dad-burn that thar fresky filly!" he cried, angrily. "Jes' brung her noisy bones up on that thar
porch agin, an' her huffs will bust spang through the
planks o' the floor the fust thing ye know."

The narrow aperture, as he held the door ajar,

showed outlined against the darkness the graceful head of a young mare; and once more hoof-beats resounded on the rotten planks of the porch.

Clouds were adrift in the sky. No star gleamed in the wide space high above the sombre mountains. On every side they encompassed Lonesome Cove, which seemed to have importunately thrust itself into the darkling solemnities of their intimacy.

All at once the ranger let the door fly from his hand, and stood gazing in blank amazement. For there was a strange motion in the void vastnesses of the wilderness. They were creeping into view. How, he could not say, but the summit of the great mountain opposite was marvellously distinct against the sky. He saw the naked, gaunt, December woods. He saw the grim, gray crags. And yet Lonesome Cove below and the spurs on the other side were all benighted. A pale, flickering light was dawning in the clouds; it brightened, faded, glowed again, and their sad, gray folds assumed a vivid vermilion reflection, for there was a fire in the forest below. Only these reactions of color on the clouds betokened its presence and its progress. Sometimes a fluctuation of orange crossed them, then a glancing line of blue, and once more that living red hue which only a pulsating flame can bestow.

" Air it the comin' o' the Jedgmint Day, Tobe?" asked his wife, in a meek whisper.

" I'd be afraid so if I war ez big a sinner ez you-uns," he returned.

" The woods air afire," the old woman declared, in a shrill voice.

"They be a-soakin' with las' night's rain," he re-torted, gruffly.

The mare was standing near the porch. Sudden-ly he mounted her and rode hastily off, without a word of his intention to the staring women in the doorway.

He left freedom of speech behind him. "Take yer bones along, then, ye tongue-tied catamount!" his wife's mother apostrophized him, with all the acrimony of long repression. "Got no mo' polite-ness 'n a settin' hen," she muttered, as she turned back into the room.

The young woman lingered wistfully. "I wisht he wouldn't go a-ridin' off that thar way 'thout lettin' we-uns know whar he air bound fur, an' when he'll kem back. He mought git hurt some ways roun' that thar fire—git overtook by it, mebbe."

"Ef he war roasted 'twould be mighty peaceful round in Lonesome," the old crone exclaimed, ran-corously.

Her daughter stood for a moment with the bar of the door in her hand, still gazing out at the flare in the sky. The unwonted emotion had conjured a change in the stereotyped patience in her face—even anxiety, even the acuteness of fear, seemed a less pathetic expression than that meek monotony bespeaking a broken spirit. As she lifted her eyes to the mountain one might wonder to see that they were so blue. In the many haggard lines drawn upon her face the effect of the straight lineaments was lost ; but just now, embellished with a flush, she looked young—as young as her years.

As she buttoned the door and put up the bar her

mother's attention was caught by the change. Peering at her critically, and shading her eyes with her hand from the uncertain flicker of the tallow dip, she broke out, passionately: "Wa'al, 'Genie, who would ever hev thought ez yer cake would be *all* dough? Sech a laffin', plump, spry gal ez ye useter be—fur all the worl' like a fresky young deer! An' sech a pack o' men ez ye hed the choice amongst! An' ter pick out Tobe Gryce an' marry him, an' kem 'way down hyar ter live along o' him in Lonesome Cove!"

She chuckled aloud, not that she relished her mirth, but the harlequinade of fate constrained a laugh for its antics. The words recalled the past to Eugenia; it rose visibly before her. She had had scant leisure to reflect that her life might have been ordered differently. In her widening eyes were new depths, a vague terror, a wild speculation, all struck aghast by its own temerity.

"Ye never said nuthin ter hender," she faltered.

"I never knowed Tobe, sca'cely. How's ennybody goin' ter know a man ez lived 'way off down hyar in Lonesome Cove?" her mother retorted, acridly, on the defensive. "He never courted *me*, nohows. All the word he gin me war, 'Howdy,' an' I gin him no less."

There was a pause.

Eugenia knelt on the hearth. She placed together the broken chunks, and fanned the flames with a turkey wing. "I won't kiver the fire yit," she said, thoughtfully. "He mought be chilled when he gits home."

The feathery flakes of the ashes flew; they caught here and there in her brown hair. The blaze flared up, and flickered over her flushed, pensive face, and glowed in her large and brilliant eyes.

"Tobe said 'Howdy,'" her mother bickered on. "I knowed by that ez he hed the gift o' speech, but he spent no mo' words on me." Then, suddenly, with a change of tone: "I war a fool, though, ter gin my cornsent ter yer marryin' him, bein' ez ye war the only child I hed, an' I knowed I'd hev ter live with ye 'way down hyar in Lonesome Cove. I wish now ez ye hed abided by yer fust choice, an' married Luke Todd."

Eugenia looked up with a gathering frown. "I hev no call ter spen' words 'bout Luke Todd," she said, with dignity, "ez me an' him are both married ter other folks."

"I never said ye hed," hastily replied the old woman, rebuked and embarrassed. Presently, however, her vagrant speculation went recklessly on. "Though ez ter Luke's marryin', 'tain't wuth while ter set store on sech. The gal he found over thar in Big Fox Valley favors ye ez close ez two black-eyed peas. That's why he married her. She looks precisely like ye useter look. An' she laffs the same. An' I reckon *she* 'ain't hed no call ter quit laffin', 'kase he air a powerful easy-goin' man. Leastways, he useter be when we-uns knowed him."

"That ain't no sign," said Eugenia. "A saafter-spoken body I never seen than Tobe war when he fust kem a-courtin' round the settlemint."

"Sech ez that ain't goin' ter las' noways," dryly remarked the philosopher of the chimney-corner.

This might seem rather a reflection upon the
courting gentry in general than a personal observation.   But Eugenia's consciousness lent it point.

"Laws-a-massy," she said, "'Tobe ain't so rampa-
gious, nohows, ez folks make him out.   He air
toler'ble peaceable, cornsiderin' ez nobody hev ever
hed grit enough ter make a stand agin him, 'thout
'twar the Cunnel thar."

She glanced around at the little girl's face framed
in the frill of her night-cap, and peaceful and infantile as it lay on the pillow.

"Whenst the Cunnel war born," Eugenia went
on, languidly reminiscent, "Tobe war powerful outed
'kase she war a gal.   I reckon ye 'members ez how
he said he hed no use for sech cattle ez that.   An'
when she tuk sick he 'lowed he seen no differ.   'Jes
ez well die. ez live,' he said.   An' bein' ailin', the
Cunnel tuk it inter her head ter holler.   Sech holler-
in' we-uns hed never hearn with none o' the t'other
chil'ren.   The boys war nowhar.   But a-fust it never
'sturbed Tobe.   He jes spoke out same ez he useter
do at the t'others, 'Shet up, ye pop-eyed buzzard!'
Wa'al, sir, the Cunnel jes blinked at him, an' braced
herself ez stiff, an' *yelled!*   I 'lowed 'twould take
off the roof.   An' Tobe said he'd wring her neck ef
she warn't so mewlin'-lookin' an' peakèd.   An' he
tuk her up an' walked across the floor with her, an'
she shet up ; an' he walked back agin, an' she stayed
shet up.   Ef he sot down fur a minit, she yelled so
ez ye'd think ye'd be deef fur life, an' ye 'most hoped
ye would be.   So Tobe war obleeged ter tote her
agin ter git shet o' the noise.   He got started on
that thar 'forced march,' ez he calls it, an' he never

could git off'n it. Trot he must when the Cunnel pleased. He 'lowed she reminded him o' that thar old Cunnel that he sarved under in the wars. Ef it killed the regiment, he got thar on time. Sence then the Cunnel jes gins Tobe her orders, an' he moseys ter do 'em quick, jes like he war obleeged ter obey. I b'lieve he air, somehows."

"Wa'al, some day," said the disaffected old woman, assuming a port of prophetic wisdom, "Tobe will find a differ. Thar ain't no man so headin' ez don't git treated with perslimness by somebody some time. I knowed a man wunst ez owned fower horses an' cattle-critters quarryspondin', an' he couldn't prove ez he war too old ter be summonsed ter work on the road, an' war fined by the overseer 'cordin' ter law. Tobe will git his wheel scotched yit, sure ez ye air born. Somebody besides the Cunnel will skeer up grit enough ter make a stand agin him. I dunno how other men kin sleep o' night, knowin' how he be always darin' folks ter differ with him, an' how brigaty he be. The Bible 'pears ter me ter hev Tobe in special mind when it gits ter mournin' 'bout'n the stiff-necked ones."

The spirited young mare that the ranger rode strove to assert herself against him now and then, as she went at a breakneck speed along the sandy bridle-path through the woods. How was she to know that the white-wanded young willow by the way-side was not some spiritual manifestation as it suddenly materialized in a broken beam from a rift in the clouds? But as she reared and plunged she felt his heavy hand and his heavy heel, and so forward again at a steady pace. The forests served to screen the strange light in the sky, and the lonely road was dark, save where the moonbeam was splintered and the mists loitered.

Presently there were cinders flying in the breeze, a smell of smoke pervaded the air, and the ranger forgot to curse the mare when she stumbled.

"I wonder," he muttered, "what them no 'count half-livers o' town folks hev hed the shiftlessness ter let ketch afire thar!"

As he neared the brink of the mountain he saw a dense column of smoke against the sky, and a break in the woods showed the little town — the few log houses, the "gyarden spots" about them, and in the centre of the Square a great mass of coals, a flame flickering here and there, and two gaunt and tottering chimneys where once the court-house had stood. At some distance — for the heat was still intense— were grouped the slouching, spiritless figures of the mountaineers. On the porches of the houses, plainly visible in the unwonted red glow, were knots of women and children — ever and anon a brat in the scantiest of raiment ran nimbly in and out. The clouds still borrowed the light from below, and the solemn, leafless woods on one side were outlined distinctly against the reflection in the sky. The flare showed, too, the abrupt precipice on the other side, the abysmal gloom of the valley, the austere summit - line of the mountain beyond, and gave the dark mysteries of the night a sombre revelation, as in visible blackness it filled the illimitable space.

The little mare was badly blown as the ranger

sprang to the ground. He himself was panting with amazement and eagerness.

"The stray-book!" he cried. "Whar's the stray-book?"

One by one the slow group turned, all looking at him with a peering expression as he loomed distorted through the shimmer of the heat above the bed of live coals and the hovering smoke.

"Whar's the stray-book?" he reiterated, imperiously.

"Whar's the court - house, I reckon ye mean to say," replied the sheriff—a burly mountaineer in brown jeans and high boots,¯ on which the spurs jingled; for in his excitement he had put them on as mechanically as his clothes, as if they were an essential part of his attire.

"Naw, I *ain't* meanin' ter say whar's the court-house," said the ranger, coming up close, with the red glow of the fire on his face, and his eyes flashing under the broad brim of his wool hat. He had a threatening aspect, and his elongated shadow, following him and repeating the menace of his attitude, seemed to back him up. "Ye air sech a triflin', slack-twisted tribe hyar in town, ez ennybody would know ef a spark cotched fire ter suthin, ye'd set an' suck yer paws, an' eye it till it bodaciously burnt up the court - house — sech a dad-burned lazy set o' half-livers ye be! I never axed 'bout'n the court-house. I want ter know whar's that thar stray-book," he concluded, inconsequently.

"Tobe Gryce, ye air fairly demented," exclaimed the register—a chin-whiskered, grizzled old fellow, sitting on a stump and hugging his knee with a desolate, bereaved look—"talkin' 'bout the *stray-book*, an' all the records gone! What will folks do 'bout thar deeds, an' mortgages, an' sech? An' that thar keerful index ez I had made—ez straight ez a string—all cinders!"

He shook his head, mourning alike for the party of the first part and the party of the second part, and the vestiges of all that they had agreed together.

"An' ye ter kem mopin' hyar this time o' night arter the *stray-book!*" said the sheriff. "Shucks!" And he turned aside and spat disdainfully on the ground.

"I want that thar stray-book!" cried Gryce, indignantly. "Ain't nobody seen it?" Then realizing the futility of the question, he yielded to a fresh burst of anger, and turned upon the bereaved register. "An' did ye jes set thar an' say, 'Good Mister Fire, don't burn the records; what 'll folks do 'bout thar deeds an' sech?' an' hold them claws o' yourn, an' see the court-house burn up, with that thar stray-book in it?"

Half a dozen men spoke up. "The fire tuk inside, an' the court-house war haffen gone 'fore 'twar seen," said one, in sulky extenuation.

"Leave Tobe be—let him jaw!" said another, cavalierly.

"Tobe 'pears ter be sp'ilin' fur a fight," said a third, impersonally, as if to direct the attention of any belligerent in the group to the opportunity.

The register had an expression of slow cunning as he cast a glance up at the overbearing ranger.

"What ailed the stray-book ter bide hyar in the

court-house all night, Tobe? Couldn't ye gin it house-room? Thar warn't no special need fur it to be hyar."

Tobe Gryce's face showed that for once he was at a loss. He glowered down at the register and said nothing.

"Ez ter me," resumed that worthy, "by the law o' the land my books war obligated ter be thar." He quoted, mournfully, "'Shall at all times be and remain in his office.'"

He gathered up his knee again and subsided into silence.

All the freakish spirits of the air were a-loose in the wind. In fitful gusts they rushed up the gorge, then suddenly the boughs would fall still again, and one could hear the eerie rout a-rioting far off down the valley. Now and then the glow of the fire would deepen, the coals tremble, and with a gleaming, fibrous swirl, like a garment of flames, a sudden animation would sweep over it, as if an apparition had passed, leaving a line of flying sparks to mark its trail.

"I'm goin' home," drawled Tobe Gryce, presently. "I don't keer a frog's toe-nail ef the whole settlemint burns bodaciously up; 'tain't nuthin ter me. I hev never hankered ter live in towns an' git tuk up with town ways, an' set an' view the court-house like the apple o' my eye. We-uns don't ketch fire down in the Cove, though mebbe we ain't so peart ez folks ez herd tergether like sheep an' sech."

The footfalls of the little black mare annotated the silence of the place as he rode away into the darkling woods. The groups gradually disappeared from the porches. The few voices that sounded at long intervals were low and drowsy. The red fire smouldered in the centre of the place, and sometimes about it appeared so doubtful a shadow that it could hardly argue substance. Far away a dog barked, and then all was still.

Presently the great mountains loom aggressively along the horizon. The black abysses, the valleys and coves, show duncolored verges and grow gradually distinct, and on the slopes the ash and the pine and the oak are all lustrous with a silver rime. The mists are rising, the wind springs up anew, the clouds set sail, and a beam slants high.

"What I want ter know," said a mountaineer newly arrived on the scene, sitting on the verge of the precipice, and dangling his long legs over the depths beneath, "air how do folks ez live 'way down in Lonesome Cove, an' who nobody knowed nuthin about noways, ever git 'lected ranger o' the county, ennyhow. I ain't s'prised none ter hear 'bout Tobe Gryce's goin's-on hyar las' night. I hev looked fur more'n that."

"Wa'al, I'll tell ye," replied the register. "Nuthin' but favoritism in the county court. Ranger air 'lected by the jestices. Ye know," he added, vainglorious of his own tenure of office by the acclaiming voice of the sovereign people, "ranger ain't 'lected, like the register, by pop'lar vote."

A slow smoke still wreathed upward from the charred ruins of the court-house. Gossiping groups stood here and there, mostly the jeans-clad mountaineers, but there were a few who wore "store

clothes," being lawyers from more sophisticated re-
gions of the circuit. Court had been in session the
previous day. The jury, serving in a criminal case
—still strictly segregated, and in charge of an officer
—were walking about wearily in double file, waiting with what patience they might their formal discharge.

The sheriff's dog, a great yellow cur, trotted in
the rear. When the officer was first elected, this
animal, observing the change in his master's habits,
deduced his own conclusions. He seemed to think
the court-house belonged to the sheriff, and thence-
forward guarded the door with snaps and growls;
being a formidable brute, his idiosyncrasies invested
the getting into and getting out of law with abnormal difficulties. Now, as he followed the disconso-
late jury, he bore the vigilant mien with which he formerly drove up the cows, and if a juror loitered
or stepped aside from the path, the dog made a
slow detour as if to round him in, and the melancholy cortége wandered on as before. More than
one looked wistfully at the group on the crag, for it
was distinguished by that sprightly interest which
scandal excites so readily.

"Ter my way of thinkin'," drawled Sam Peters,
swinging his feet over the giddy depths of the
valley, "Tobe ain't sech ez oughter be set over
the county ez a ranger, noways. 'Pears not ter
me, an' I hev been keepin' my eye on him mighty
sharp."

A shadow fell among the group, and a man sat
down on a bowlder hard by. He, too, had just
arrived, being lured to the town by the news of the

fire. His slide had been left at the verge of the clearing, and one of the oxen had already lain down; the other, although hampered by the yoke thus diagonally displaced, stood meditatively gazing at the distant blue mountains. Their master nodded a slow, grave salutation to the group, produced a plug of tobacco, gnawed a fragment from it, and restored it to his pocket. He had a pensive face, with an expression which in a man of wider culture we should discriminate as denoting sensibility. He had long yellow hair that hung down to his shoulders, and a tangled yellow beard. There was something at once wistful and searching in his gray eyes, dull enough, too, at times. He lifted them heavily, and they had a drooping lid and lash. There seemed an odd incongruity between this sensitive, weary face and his stalwart physique. He was tall and well proportioned. A leather belt girded his brown jeans coat. His great cowhide boots were drawn to the knee over his trousers. His pose, as he leaned on the rock, had a muscular picturesqueness.

"Who be ye a-talkin' about?" he drawled.

Peters relished his opportunity. He laughed in a distorted fashion, his pipe-stem held between his teeth.

"*You-uns* ain't wantin' ter swop lies 'bout sech ez him, Luke! We war a-talkin' 'bout Tobe Gryce."

The color flared into the new-comer's face. A sudden animation fired his eye.

"Tobe Gryce air jes the man I'm always wantin' ter hear a word about. Jes perceed with yer rat-killin'. I'm with ye." And Luke Todd placed his

elbows on his knees and leaned forward with an air
of attention.

Peters looked at him, hardly comprehending this
ebullition. It was not what he had expected to elicit.
No one laughed. His fleer was wide of the mark.

"Wa'al" — he made another effort — "Tobe, we
war jes sayin', ain't fitten fur ter be ranger o' the
county. He be ez peart in gittin' ter own other
folkses' stray cattle ez he war in courtin' other
folkses' sweetheart, an', ef the truth mus' be knowed,
in marryin' her." He suddenly twisted round, in
some danger of falling from his perch. "I want
ter ax one o' them thar big - headed lawyers a question on a p'int o' law," he broke off, abruptly.

"What be 'Tobe Gryce a-doin' of now?" asked
Luke Todd, with eager interest in the subject.

"Wa'al," resumed Peters, nowise loath to return
to the gossip, "Tobe, ye see, air the ranger o' this
hyar county, an' by law all the stray horses ez air
tuk up by folks hev ter be reported ter him, an' appraised by two householders, an' swore to afore the
magistrate an' be advertised by the ranger, an' ef
they ain't claimed 'fore twelve months, the taker-up
kin pay into the county treasury one-haffen the appraisement an' hev the critter fur his'n. An' the
owner can't prove it away arter that."

"Thanky," said Luke Todd, dryly. "S'pose ye
teach yer gran'mammy ter suck aigs. I knowed all
that afore."

Peters was abashed, and with some difficulty collected himself.

"An' I knowed ye knowed it, Luke," he hastily
conceded. "But hyar be what I'm a-lookin' at—

the law 'ain't got no pervision fur a stray horse ez
kem of a dark night, 'thout nobody's percuremint,
ter the ranger's own house.   Now, the p'int o' law
ez I wanted ter ax the lawyers 'bout air this — kin
the ranger be the ranger an' the taker-up too?"

He turned his eyes upon the great landscape lying
beneath, flooded with the chill matutinal sunshine,
and flecked here and there with the elusive shadows
of the fleecy drifting clouds.   Far away the long
horizontal lines of the wooded spurs, converging on
either side of the valley and rising one behind the
other, wore a subdued azure, all unlike the burning
blue of summer, and lay along the calm, passionless
sky, that itself was of a dim, repressed tone.   On the
slopes nearer, the leafless boughs, massed together,
had purplish‑garnet depths of color wherever the
sunshine struck aslant, and showed richly against
the faintly tinted horizon.   Here and there among
the boldly jutting gray crags hung an evergreen-vine,
and from a gorge on the opposite mountain gleamed
a continuous flash, like the waving of a silver plume,
where a cataract sprang down the rocks.   In the
depths of the valley, a field in which crab-grass had
grown in the place of the harvested wheat showed a
tiny square of palest yellow, and beside it a red clay
road, running over a hill, was visible.   Above all a
hawk was flying.

" Afore the winter fairly set in las' year," Peters
resumed, presently, "a stray kem ter Tobe's house.
He 'lowed ter me ez he fund her a-standin' by the
fodder-stack a-pullin' off'n it.   An' he 'quired round,
an' he never hearn o' no owner.   I reckon he never
axed outside o' Lonesome," he added, cynically.

He puffed industriously at his pipe for a few moments; then continued : " Wa'al, he 'lowed he couldn't feed the critter fur fun.   An' he couldn't work her till she war appraised an' sech, that bein' agin the law fur strays.   So he jes ondertook ter be ranger an' taker-up too — the bangedest consarn in the kentry !   Ef the leetle mare hed been wall-eyed, or lame, or ennything, he wouldn't hev wanted ter be ranger an' taker-up too.   But she air the peartest little beastis—she war jes bridle-wise when she fust kem—young an' spry !"

Luke Todd was about to ask a question, but Peters, disregarding him, persisted :

"Wa'al, Tobe tuk up the beastis, an' I reckon he reported her ter hisself, bein' the ranger—the critter makes me laff—an' he hed that thar old haffen-blind uncle o' his'n an' Perkins Bates, ez be never sober, ter appraise the vally o' the mare, an' I s'pose he delivered thar certificate ter hisself, an' I reckon he tuk oath that she kem 'thout his procure*mint* ter his place, in the presence o' the ranger."

"I reckon thar ain't no law agin the ranger's bein' a ranger an' a taker-up too," put in one of the bystanders.   " 'Tain't like a sher'ff's buyin' at his own sale.   An' he hed ter pay haffen her vally into the treasury o' the county arter twelve months, ef the owner never proved her away."

"Thar ain't no sign he ever paid a cent," said Peters, with a malicious grin, pointing at the charred remains of the court-house, " an' the treasurer air jes dead."

" Wa'al, Tobe hed ter make a report ter the jedge o' the county court every six months."

"'The papers of his office air cinders," retorted Peters.

"Wa'al, then," argued the optimist, "the stray-book will show ez she war reported an' sech."

"The ranger took mighty partic'lar pains ter hev his stray-book in that thar court-house when 'twar burnt."

There was a long pause while the party sat ruminating upon the suspicions thus suggested.

Luke Todd heard them, not without a thrill of satisfaction. He found them easy to adopt. And he, too, had a disposition to theorize.

"It takes a mighty mean man ter steal a horse," he said. "Stealin' a horse air powerful close ter murder. Folkses' lives fairly depend on a horse ter work thar corn an' sech, an' make a support fur em. I hev knowed folks ter kem mighty close ter starvin' through hevin thar horse stole. Why, even that thar leetle filly of our'n, though she hedn't been fairly bruk ter the plough, war mightily missed. We-uns hed ter make out with the old sorrel, ez air nigh fourteen year old, ter work the crap, an' we war powerful disapp'inted. But we ain't never fund no trace o' the filly sence she war tolled off one night las' fall a year ago."

The hawk floating above the valley and its winged shadow disappeared together in the dense glooms of a deep gorge. Luke Todd watched them as they vanished.

Suddenly he lifted his eyes. They were wide with a new speculation. An angry flare blazed in them. "What sort'n beastis is this hyar mare ez the ranger tuk up?" he asked.

Peters looked at him, hardly comprehending his

tremor of excitement. "Seems sorter sizable," he replied, sibilantly, sucking his pipe-stem.

Todd nodded meditatively several times, leaning his elbows on his knees, his eyes fixed on the landscape. "Hev she got enny partic'lar marks, ez ye knows on?" he drawled.

"Wa'al, she be ez black ez a crow, with the nigh fore-foot white. An' she hev got a white star spang in the middle o' her forehead, an' the left side o' her nose is white too."

Todd rose suddenly to his feet. "By gum!" he cried, with a burst of passion, "she air *my* filly! An' 'twar that thar durned horse-thief of a ranger ez tolled her off!"

Deep among the wooded spurs Lonesome Cove nestles, sequestered from the world. Naught emigrates thence except an importunate stream that forces its way through a rocky gap, and so to freedom beyond. No stranger intrudes ; only the moon looks in once in a while. The roaming wind may explore its solitudes ; and it is but the vertical sunbeams that strike to the heart of the little basin, because of the massive mountains that wall it round and serve to isolate it. So nearly do they meet at the gap that one great assertive crag, beetling far above, intercepts the view of the wide landscape beyond, leaving its substituted profile jaggedly serrating the changing sky. Above it, when the weather is fair, appear vague blue lines, distant mountain summits, cloud strata, visions. Below its jutting verge may be caught glimpses of the widening valley without. But pre-eminent, gaunt, sombre, it sternly dominates

"Lonesome," and is the salient feature of the little world it limits.

Tobe Gryce's house, gray, weather - beaten, moss-grown, had in comparison an ephemeral, modern aspect. For a hundred years its inmates had come and gone and lived and died. They took no heed of the crag, but never a sound was lost upon it. Their drawling iterative speech the iterative echoes conned. The ringing blast of a horn set astir some phantom chase in the air. When the cows came lowing home, there were lowing herds in viewless company. Even if one of the children sat on a rotting log crooning a vague, fragmentary ditty, some faint - voiced spirit in the rock would sing. Lonesome Cove?—home of invisible throngs!

As the ranger trotted down the winding road, multitudinous hoof - beats, as of a troop of cavalry, heralded his approach to the little girl who stood on the porch of the log-cabin and watched for him.

"Hy're, Cunnel!" he cried, cordially.

But the little "Colonel" took no heed. She looked beyond him at the vague blue mountains, against which the great grim rock was heavily imposed, every ledge, every waving dead crisp weed, distinct.

He noticed the smoke curling briskly up in the sunshine from the clay and stick chimney. He strode past her into the house, as Eugenia, with all semblance of youth faded from her countenance, haggard and hollow-eyed in the morning light, was hurrying the corn-dodgers and venison steak on the table.

Perhaps he did not appreciate that the women were pining with curiosity, for he vouchsafed no word of the excitements in the little town ; and he himself was ill at ease.

"What ails the Cunnel, 'Genie?" he asked, presently, glancing up sharply from under his hat brim, and speaking with his mouth full.

"The cat 'pears ter hev got her tongue," said Eugenia, intending that the "Colonel" should hear, and perhaps profit. "She ain't able ter talk none this mornin'." .

The little body cast so frowning a glance upon them as she stood in the doorway that her expression was but slightly less lowering than her father's. It was an incongruous demonstration, with her infantile features, her little yellow head, and the slight physical force she represented. She wore a blue cotton frock, fastened up the back with great horn buttons ; she had on shoes laced with leather strings ; one of her blue woollen stockings fell over her ankle, disclosing the pinkest of plump calves; the other stocking was held in place by an unabashed cotton string. She had a light in her dark eyes and a color in her cheek, and albeit so slight a thing, she wielded a strong coercion.

"Laws-a-massy, Cunnel!" said Tobe, in a harried manner, "couldn't ye find me nowhar? I'm powerful sorry. I couldn't git back hyar no sooner."

But not in this wise was she to be placated. She fixed her eyes upon him, but made no sign.

He suddenly rose from his half-finished breakfast. "Look-a-hyar, Cunnel," he cried, joyously, "don't ye want ter ride the filly?—ye knew ye hanker ter ride the filly."

Even then she tried to frown, but the bliss of the

prospect overbore her. Her cheek and chin dimpled, and there was a gurgling display of two rows of jagged little teeth as the doughty "Colonel" was swung to his shoulder and he stepped out of the door.

He laughed as he stood by the glossy black mare and lifted the child to the saddle. The animal arched her neck and turned her head and gazed back at him curiously. "Hold on tight, Cunnel," he said as he looked up at her, his face strangely softened almost beyond recognition. And she gurgled and laughed and screamed with delight as he began to slowly lead the mare along.

The "Colonel" had the gift of continuance. Some time elapsed before she exhausted the joys of exaltation. More than once she absolutely refused to dismount. Tobe patiently led the beast up and down, and the "Colonel" rode in state. It was only when the sun had grown high, and occasionally she was fain to lift her chubby hands to her eyes, imperiling her safety on the saddle, that he ventured to seriously remonstrate, and finally she permitted herself to be assisted to the ground. When, with the little girl at his heels, he reached the porch, he took off his hat, and wiped the perspiration from his brow with his great brown hand.

"I tell ye, jouncin' round arter the Cunnel air powerful hot work," he declared.

The next moment he paused. His wife had come to the door, and there was a strange expression of alarm among the anxious lines of her face.

"Tobe," she said, in a bated voice, "who war them men?"

He stared at her, whirled about, surveyed the vacant landscape, and once more turned dumfounded toward her. "What men?" he asked.

"Them men ez acted so cur'ous," she said. "I couldn't see thar faces plain, an' I dunno who they war."

"Whar war they?" And he looked over his shoulder once more.

"Yander along the ledges of the big rock. Thar war two of 'em, hidin' ahint that thar jagged aidge. An' ef yer back war turned they'd peep out at ye an' the Cunnel ridin'. But whenst ye would face round agin, they'd drap down ahint the aidge o' the rock. I 'lowed wunst ez I'd holler ter ye, but I war feared ye moughtn't keer ter know." Her voice fell in its deprecatory cadence.

He stood in silent perplexity. "Ye air a fool, 'Genie, an' ye never seen nuthin'. Nobody hev got enny call ter spy on me."

He stepped in-doors, took down his rifle from the rack, and went out frowning into the sunlight.

The suggestion of mystery angered him. He had a vague sense of impending danger. As he made his way along the slope toward the great beetling crag all his faculties were on the alert. He saw naught unusual when he stood upon its dark-seamed summit, and he went cautiously to the verge and looked down at the many ledges. They jutted out at irregular intervals, the first only six feet below, and all accessible enough to an expert climber. A bush grew in a niche. An empty nest, riddled by the wind, hung dishevelled from a twig. Coarse withered grass tufted the crevices.

Far below he saw the depths of the Cove—the tops of the leafless trees, and, glimpsed through the interlacing boughs, the rush of a mountain rill, and a white flash as a sunbeam slanted on the foam.

He was turning away, all incredulous, when with a sudden start he looked back. On one of the ledges was a slight depression. It was filled with sand and earth. Imprinted upon it was the shape of a man's foot. The ranger paused and gazed fixedly at it. "Wa'al, by the Lord!" he exclaimed, under his breath. Presently, "But they hev no call!" he argued. Then once more, softly, "By the Lord!"

The mystery baffled him. More than once that day he went up to the crag and stood and stared futilely at the footprint. Conjecture had license and limitations, too. As the hours wore on he became harassed by the sense of espionage. He was a bold man before the foes he knew, but this idea of inimical lurking, of furtive scrutiny for unknown purposes, preyed upon him. He brooded over it as he sat idle by the fire. Once he went to the door and stared speculatively at the great profile of the cliff. The sky above it was all a lustrous amber, for the early sunset of the shortest days of the year was at hand. The mountains, seen partly above and partly below it, wore a glamourous purple. There were clouds, and from their rifts long divergent lines of light slanted down upon the valley, distinct among their shadows. The sun was not visible—only in the western heavens was a half-veiled effulgence too dazzlingly white to be gazed upon. The ranger shaded his eyes with his hand.

No motion, no sound; for the first time in his life the unutterable loneliness of the place impressed him.

"'Genie," he said, suddenly, looking over his shoulder within the cabin, "be you-uns *sure* ez they war— *folks?*"

"I dunno what you mean," she faltered, her eyes dilated. "They *looked* like folks."

"I reckon they war," he said, reassuring himself. "The Lord knows I hope they war."

That night the wind rose. The stars all seemed to have burst from their moorings, and were wildly adrift in the sky. There was a broken tumult of billowy clouds, and the moon tossed hopelessly amongst them, a lunar wreck, sometimes on her beam ends, sometimes half submerged, once more gallantly struggling to the surface, and again sunk. The bare boughs of the trees beat together in a dirge-like monotone. Now and again a leaf went sibilantly whistling past. The wild commotion of the heavens and earth was visible, for the night was not dark. The ranger, standing within the rude stable of unhewn logs, all undaubed, noted how pale were the horizontal bars of gray light alternating with the black logs of the wall. He was giving the mare a feed of corn, but he had not brought his lantern, as was his custom. That mysterious espionage had in some sort shaken his courage, and he felt the obscurity a shield. He had brought, instead, his rifle.

The equine form was barely visible among the glooms. Now and then, as the mare noisily munched,

she lifted a hoof and struck it upon the ground
with a dull thud.   How the gusts outside were swirling up the gorge!   The pines swayed and sighed.
Again the boughs of the chestnut-oak above the
roof crashed together.   Did a fitful blast stir the
door?

He lifted his eyes mechanically.   A cold thrill
ran through every fibre.   For there, close by the
door, somebody—something—was peering through
the space between the logs of the wall.   The face
was invisible, but the shape of a man's head was
distinctly defined.   He realized that it was no
supernatural manifestation when a husky voice began to call the mare, in a hoarse whisper, "Cobe!
Cobe! Cobe!"   With a galvanic start he was about
to spring forward to hold the door.   A hand from
without was laid upon it.

He placed the muzzle of his gun between the
logs, a jet of red light was suddenly projected into
the darkness, the mare was rearing and plunging
violently, the little shanty was surcharged with roar
and reverberation, and far and wide the crags and
chasms echoed the report of the rifle.

There was a vague clamor outside, an oath, a cry
of pain.   Hasty footfalls sounded among the dead
leaves and died in the distance.

When the ranger ventured out he saw the door of
his house wide open, and the firelight flickering out
among the leafless bushes.   His wife met him half-
way down the hill.

"Air ye hurt, Tobe?" she cried.   "Did yer gun
go off suddint?"

"Mighty suddint," he replied, savagely.

"Ye didn't fire it a-purpose?" she faltered.

"Edzactly so," he declared.

"Ye never hurt nobody, did ye, Tobe?" She had turned very pale. "I 'lowed it couldn't be the wind ez I hearn a-hollerin'."

"I hopes an' prays I hurt 'em," he said, as he re-placed the rifle in the rack. He was shaking the other hand, which had been jarred in some way by the hasty discharge of the weapon. "Some dad-burned horse-thief war arter the mare. Jedgin' from the sound o' thar runnin', 'peared like to me ez thar mought be two o' 'em."

The next day the mare disappeared from the stable. Yet she could not be far off, for Tobe was about the house most of the time, and when he and the "Colonel" came in-doors in the evening the little girl held in her hand a half-munched ear of corn, evidently abstracted from the mare's supper.

"Whar be the filly hid, Tobe?" Eugenia asked, curiosity overpowering her.

"Ax me no questions an' I'll tell ye no lies," he replied, gruffly.

In the morning there was a fall of snow, and she had some doubt whether her mother, who had gone several days before to a neighbor's on the summit of the range, would return; but presently the creak of unoiled axles heralded the approach of a wagon, and soon the old woman, bundled in shawls, was sitting by the fire. She wore heavy woollen socks over her shoes as protection against the snow. The incom-patibility of the shape of the hose with the human foot was rather marked, and as they were somewhat inelastic as well, there was a muscular struggle to get them off only exceeded by the effort which had been required to get them on. She shook her head again and again, with a red face, as she bent over the socks, but plainly more than this discomfort vexed her.

"Laws-a-massy, 'Genie! I hearn a awful tale over yander 'mongst them Jenkins folks. Ye oughter hev married Luke Todd, an' so I tole ye an' fairly beset ye ter do ten year ago. *He* keered fur ye. An' Tobe—shucks! Wa'al, laws-a-massy, child! I hearn a awful tale 'bout Tobe up yander at Jenkinses'."

Eugenia colored.

"Folks hed better take keer how they talk 'bout Tobe," she said, with a touch of pride. "They be powerful keerful ter do it out'n rifle range."

With one more mighty tug the sock came off, the red face was lifted, and Mrs. Pearce shook her head ruefully.

"The Bible say 'words air foolishness.' Ye dunno what ye air talkin' 'bout, child."

With this melancholy preamble she detailed the gossip that had arisen at the county town and pervaded the country-side. Eugenia commented, denied, flashed into rage, then lapsed into silence. Although it did not constrain credulity, there was something that made her afraid when her mother said :

"Ye hed better not be talkin' 'bout rifle range so brash, 'Genie, nohows. They 'lowed ez Luke Todd an' Sam Peters kem hyar—'twar jes night before las' — aimin' ter take the mare away 'thout no words an' no lawin', 'kase they didn't want ter wait. Luke hed got a chance ter view the mare, an' knowed

11

ez she war his'n.   An' Tobe war hid in the dark beside the mare, an' fired at 'em, an' the rifle-ball tuk
Sam right through the beam o' his arm.   I reckon, though, ez that warn't true, else ye would hev
knowed it."

She looked up anxiously over her spectacles at
her daughter.

"I hearn Tobe shoot," faltered Eugenia.   "I seen
blood on the leaves."

"Laws-a-massy!" exclaimed the old woman, irritably.   "I be fairly feared ter bide hyar ; 'twouldn't
s'prise me none ef they kem hyar an' hauled Tobe
out an' lynched him an' sech, an' who knows who
mought git hurt in the scrimmage?"

They both fell silent as the ranger strode in.
They would need a braver heart than either bore to
reveal to him the suspicions of horse - stealing sown
broadcast over the mountain.   Eugenia felt that
this in itself was coercive evidence of his innocence.
Who dared so much as say a word to his face?

The weight of the secret asserted itself, however.
As she went about her accustomed tasks, all bereft
of their wonted interest, vapid and burdensome, she
carried so woe-begone a face that it caught his attention, and he demanded, angrily,

"What ails ye ter look so durned peakèd?"

This did not abide long in his memory, however,
and it cost her a pang to see him so unconscious.

She went out upon the porch late that afternoon
to judge of the weather.   Snow was falling again.
The distant summits had disappeared.   The mountains near at hand loomed through the myriads of
serried white flakes.   A crow flew across the Cove

in its midst. It heavily thatched the cabin, and
tufts dislodged by the opening of the door fell down
upon her hair. Drifts lay about the porch. Each
rail of the fence was laden. The ground, the rocks,
were deeply covered. She reflected with satisfaction
that the red splotch of blood on the dead leaves was
no longer visible. Then a sudden idea struck her
that took her breath away. She came in, her cheeks
flushed, her eyes bright, with an excited dubitation.

Her husband commented on the change. " Ye
air a powerful cur'ous critter, 'Genie," he said: "a
while ago ye looked some fower or five hundred
year old — now ye favors yerself when I fust kem
a-courtin' round the settlemint."

She hardly knew whether the dull stir in her heart
were pleasure or pain. Her eyes filled with tears,
and the irradiated iris shone through them with a
liquid lustre. She could not speak.

Her mother took ephemeral advantage of his
softening mood. " Ye useter be mighty perlite and
saaft-spoken in them days, Tobe," she ventured.

" I hed ter be," he admitted, frankly, " 'kase thar
war sech a many o' them mealy - mouthed cusses a-
waitin' on 'Genie. The kentry 'peared ter me ter
bristle with Luke Todd ; he 'minded me o' brum-
saidge—*everywhar* ye seen his yaller head, ez homely
an' ez onwelcome."

" I never wunst gin Luke a thought arter ye tuk
ter comin' round the settlemint," Eugenia said, softly.

" I wisht I hed knowed that then," he replied ;
"else I wouldn't hev been so all-fired oneasy an'
beset. I wasted mo' time a-studyin' 'bout ye an'
Luke Todd 'n ye war both wuth, an' went 'thout

my vittles an' sot up o' nights.   Ef I hed spent that
time a-moanin' fur my sins an' settin' my soul at
peace, I'd be 'quirin' roun' the throne o' Grace now !
Young folks air powerful fursaken fools."

Somehow her heart was warmer for this allusion.
She was more hopeful.   Her resolve grew stronger
and stronger as she sat and knitted, and looked at
the fire and saw among the coals all her old life at
the settlement newly aglow.   She was remembering
now that Luke Todd had been as wax in her hands.
She recalled that when she was married there was a
gleeful " sayin' " going the rounds of the mountain
that he had taken to the woods with grief, and he was
heard of no more for weeks.   The gossips relished
his despair as the corollary of the happy bridal.   He
had had no reproaches for her.   He had only looked
the other way when they met, and she had not spoken to him since.

"He set store by my word in them days," she
said to herself, her lips vaguely moving.   " I misdoubts ef he hev furgot."

All through the long hours of the winter night she
silently canvassed her plan.   The house was still
noiseless and dark when she softly opened the door
and softly closed it behind her.

It had ceased to snow, and the sky had cleared.
The trees, all the limbs whitened, were outlined distinctly upon it, and through the boughs overhead a
brilliant star, aloof and splendid, looked coldly down.
Along dark spaces Orion had drawn his glittering
blade.   Above the snowy mountains a melancholy
waning moon was swinging.   The valley was full of
mist, white and shining where the light fell upon it,

a vaporous purple where the shadows held sway.
So still it was! the only motion in all the world the
throbbing stars and her palpitating heart.   So solemnly silent!   It was a relief, as she trudged on
and on, to note a gradual change ; to watch the sky
withdraw, seeming fainter ; to see the moon grow
filmy, like some figment of the frost ; to mark the
gray mist steal on apace, wrap mountain, valley,
and heaven with mystic folds, shut out all vision of
things familiar.   Through it only the sense of dawn
could creep.

She recognized the locality ; her breath was short ;
her step quickened.   She appeared, like an apparition out of the mists, close to a fence, and peered
through the snow - laden rails.   A sudden pang
pierced her heart.

For there, within the enclosure, milking the cow,
she saw, all blooming in the snow—herself ; the
azalea-like girl she had been!

She had not known how dear to her was that
bright young identity she remembered.   She had
not realized how far it had gone from her.   She
felt a forlorn changeling looking upon her own es-
tranged estate.

A faint cry escaped her.

The cow, with lifted head and a muttered low of
surprise, moved out of reach of the milker, who,
half kneeling upon the ground, stared with wide
blue eyes at her ghost in the mist.

There was a pause.   It was only a moment before Eugenia spoke ; it seemed years, so charged
it was with retrospect.

"I kem over hyar ter hev a word with ye," she said.

At the sound of a human voice Luke Todd's wife struggled to her feet. She held the piggin with one arm encircled about it, and with the other hand she clutched the plaid shawl around her throat. Her bright hair was tossed by the rising wind.

"I 'lowed I'd find ye hyar a-milkin' 'bout now."

The homely allusion reassured the younger woman.

"I hev ter begin toler'ble early," she said. "Spot gins 'bout a gallon a milkin' now."

Spot's calf, which subsisted on what was left over, seemed to find it cruel that delay should be added to his hardships, and he lifted up his voice in a plaintive remonstrance. This reminded Mrs. Todd of his existence; she turned and let down the bars that served to exclude him.

The stranger was staring at her very hard. Somehow she quailed under that look. Though it was fixed upon her in unvarying intensity, it had a strange impersonality. This woman was not seeing her, despite that wide, wistful, yearning gaze; she was thinking of something else, seeing some one else.

And suddenly Luke Todd's wife began to stare at the visitor very hard, and to think of something that was not before her.

"I be the ranger's wife," said Eugenia. "I kem over hyar ter tell ye he never tuk yer black mare nowise but honest, bein' the ranger."

She found it difficult to say more. Under that speculative, unseeing look she too faltered.

"They tell me ez Luke Todd air powerful outed

'bout'n it. An' I 'lowed ef he knowed from me ez
'twar tuk fair, he'd b'lieve me."

She hesitated. Her courage was flagging; her
hope had fled. The eyes of the man's wife burned
upon her face. .

"We-uns useter be toler'ble well 'quainted 'fore
he ever seen ye, an' I 'lowed he'd b'lieve my word,"
Eugenia continued.

Another silence. The sun was rising; long li-
quescent lines of light of purest amber-color were
streaming through the snowy woods; the shadows
of the fence rails alternated with bars of dazzling
glister; elusive prismatic gleams of rose and lilac
and blue shimmered on every slope — thus the
winter flowered. Tiny snow-birds were hopping
about; a great dog came down from the little snow-
thatched cabin, and was stretching himself elastically and yawning most portentously.

"An' I 'lowed I'd see ye an' git you-uns ter tell
him that word from me, an' then he'd b'lieve it,"
said Eugenia.

The younger woman nodded mechanically, still
gazing at her.

And was this her mission! Somehow it had lost
its urgency. Where was its potency, her enthusiasm? Eugenia realized that her feet were wet, her
skirts draggled; that she was chilled to the bone
and trembling violently. She looked about her
doubtfully. Then her eyes came back to the face
of the woman before her.

"Ye'll tell him, I s'pose?"

Once more Luke Todd's wife nodded mechanically, still staring.

There was nothing further to be said. A vacant interval ensued. Then, "I 'lowed I'd tell ye," Eugenia reiterated, vaguely, and turned away, vanishing with the vanishing mists.

Luke Todd's wife stood gazing at the fence through which the apparition had peered. She could see yet her own face there, grown old and worn. The dog wagged his tail and pressed against her, looking up and claiming her notice. Once more he stretched himself elastically and yawned widely, with shrill variations of tone. The calf was frisking about in awkward bovine elation, and now and then the cow affectionately licked its coat with the air of making its toilet. An assertive chanticleer was proclaiming the dawn within the hen-house, whence came too an impatient clamor, for the door, which served to exclude any marauding fox, was still closed upon the imprisoned poultry. Still she looked steadily at the fence where the ranger's wife had stood.

"That thar woman favors me," she said, presently. And suddenly she burst into tears.

Perhaps it was well that Eugenia could not see Luke Todd's expression as his wife recounted the scene. She gave it truly, but without, alas! the glamour of sympathy.

"She 'lowed ez ye'd b'lieve her, bein' ez ye useter be 'quainted."

His face flushed. "Wa'al, sir! the insurance o' that thar woman!" he exclaimed. "I war 'quainted with her; I war mighty well 'quainted with her." He had a casual remembrance of those days when "he tuk ter the woods ter wear out his grief."

"She never gin me no promise, but me an' her war courtin' some. Sech dependence ez I put on her war mightily wasted. I dunno what ails the critter ter 'low ez I set store by her word."

Poor Eugenia! There is nothing so dead as ashes. His flame had clean burned out. So far afield were all his thoughts that he stood amazed when his wife, with a sudden burst of tears, declared passionately that she knew it—she saw it—she favored Eugenia Gryce. She had found out that he had married her because she looked like another woman.

"'Genie Gryce hev got powerful little ter do ter kem a-jouncin' through the snow over hyar ter try ter set ye an' me agin one another," he exclaimed, angrily. "Stealin' the filly ain't enough ter sati'fy her!"

His wife was in some sort mollified. She sought to reassure herself.

"Air we-uns of a favor?"

"I dunno," he replied, sulkily. "I 'ain't seen the critter fur nigh on ter ten year. I hev furgot the looks of her. 'Pears like ter me," he went on, ruminating, "ez 'twar in my mind when I fust seen ye ez thar war a favor 'twixt ye. But I misdoubts now. Do she 'low ez I hev hed nuthin ter study 'bout sence?"

Perhaps Eugenia is not the only woman who overrates the strength of a sentimental attachment. A gloomy intuition of failure kept her company all the lengthening way home. The chill splendors of the wintry day grated upon her dreary mood. How should she care for the depth and richness of the

blue deepening toward the zenith in those vast
skies? What was it to her that the dead vines,
climbing the grim rugged crags, were laden with
tufts and corollated shapes wherever these fantasies
of flowers might cling, or that the snow flashed with
crystalline scintillations? She only knew that they
glimmered and dazzled upon the tears in her eyes,
and she was moved to shed them afresh. She did
not wonder whether her venture had resulted amiss.
She only wondered that she had tried aught. And
she was humbled.

When she reached Lonesome Cove she found the
piggin where she had hid it, and milked the cow
in haste. It was no great task, for the animal was
going dry. "Their'n gins a gallon a milkin'," she
said, in rueful comparison.

As she came up the slope with the piggin on her
head, her husband was looking down from the
porch with a lowering brow. "Why n't ye spen'
the day a-milkin' the cow?" he drawled. "Dawdlin'
yander in the cow-pen till this time in the mornin'!
An' ter-morrer's Chrismus!"

The word smote upon her weary heart with a
dull pain. She had no cultured phrase to char-
acterize the sensation as a presentiment, but she
was conscious of the prophetic process. To-night
"all the mounting" would be riotous with that du-
bious hilarity known as "Chrismus in the bones,"
and there was no telling what might come from the
combined orgy and an inflamed public spirit.

She remembered the familiar doom of the mountain horse-thief, the men lurking on the cliff, the
inimical feeling against the ranger. She furtively

watched him with forebodings as he came and went at intervals throughout the day.

Dusk had fallen when he suddenly looked in and beckoned to the "Colonel," who required him to take her with him whenever he fed the mare.

"Let me tie this hyar comforter over the Cunnel's head," Eugenia said, as he bundled the child in a shawl and lifted her in his arms.

"'Tain't no use," he declared. "The Cunnel ain't travellin' fur."

She heard him step from the creaking porch. She heard the dreary wind without.

Within, the clumsy shadows of the warping - bars, the spinning - wheel, and the churn were dancing in the firelight on the wall. The supper was cooking on the live coals. The children, popping corn in the ashes, were laughing ; as her eye fell upon the "Colonel's" vacant little chair her mind returned to the child's excursion with her father, and again she wondered futilely where the mare could be hid. The next moment she was heartily glad that she did not know.

It was like the fulfillment of some dreadful dream when the door opened. A man entered softly, slowly ; the flickering fire showed his shadow—was it ?—nay, another man, and still another, and another.

The old crone in the corner sprang up, screaming in a shrill, tremulous, cracked voice. For they were masked. Over the face of each dangled a bit of homespun, with great empty sockets through which eyes vaguely glanced. Even the coarse fibre of the intruders responded to that quavering, thrilling appeal. One spoke instantly :

"Laws-a-massy! Mis' Pearce, don't ye feel inter-rupted none—nor Mis' Gryce nuther. We-uns ain't harmful noways — jes want ter know whar that thar black mare hev disappeared to. She ain't in the barn."

He turned his great eye-sockets on Eugenia. The plaid homespun mask dangling about his face was grotesquely incongruous with his intent, serious gaze.

"I dunno," she faltered; "I dunno."

She had caught at the spinning-wheel for support. The fire crackled. The baby was counting aloud the grains of corn popping from the ashes. "Six, two, free," he babbled. The kettle merrily sang.

The man still stared silently at the ranger's wife. The expression in his eyes changed suddenly. He chuckled derisively. The others echoed his mock-ing mirth. "Ha! ha! ha!" they laughed aloud; and the eye-sockets in the homespun masks all glared significantly at each other. Even the dog detected something sinister in this laughter. He had been sniffing about the heels of the strangers; he bristled now, showed his teeth, and growled. The spokesman hastily kicked him in the ribs, and the animal fled yelping to the farther side of the fire-place behind the baby, where he stood and barked defiance. The rafters rang with the sound.

Some one on the porch without spoke to the leader in a low voice. This man, who seemed to have a desire to conceal his identity which could not be served by a mask, held the door with one hand that the wind might not blow it wide open. The draught fanned the fire. Once the great bowing, waving

white blaze sent a long, quivering line of light
through the narrow aperture, and Eugenia saw the
dark lurking figure outside. He had one arm in
a sling. She needed no confirmation to assure her
that this was Sam Peters, whom her husband had
shot at the stable door.

The leader instantly accepted his suggestion.
"Wa'al, Mis' Gryce, I reckon ye dunno whar Tobe
be, nuther?"

"Naw, I dunno," she said, in a tremor.

The homespun mask swayed with the distortions
of his face as he sneered:

"Ye mean ter say ye don't 'low ter tell us."

"I dunno whar he be." Her voice had sunk to
a whisper.

Another exchange of glances.

"Wa'al, ma'am, jes gin us the favor of a light by
yer fire, an' we-uns 'll find him."

He stepped swiftly forward, thrust a pine torch
into the coals, and with it all whitely flaring ran out
into the night; the others followed his example;
and the terror-stricken women, hastily barring up
the door, peered after them through the little batten
shutter of the window.

The torches were already scattered about the
slopes of Lonesome Cove like a fallen constellation.
What shafts of white light they cast upon the snow
in the midst of the dense blackness of the night!
Somehow they seemed endowed with volition, as
they moved hither and thither, for their brilliancy
almost cancelled the figures of the men that bore
them—only an occasional erratic shapeless shadow

was visible. Now and then a flare pierced the icicle-
tipped holly bushes, and again there was a fibrous
glimmer in the fringed pines.

The search was terribly silent. The snow deadened the tread. Only the wind was loud among the
muffled trees, and sometimes a dull thud sounded
when the weight of snow fell from the evergreen
laurel as the men thrashed through its dense growth.
They separated after a time, and only here and
there an isolated stellular light illumined the snow,
and conjured white mystic circles into the wide
spaces of the darkness. The effort flagged at last,
and its futility sharpened the sense of injury in Luke
Todd's heart.

He was alone now, close upon the great rock, and
looking at its jagged ledges all cloaked with snow.
Above those soft white outlines drawn against the
deep clear sky the frosty stars scintillated. Beneath
were the abysmal depths of the valley masked by the
darkness.

His pride was touched. In the old quarrel his
revenge had been hampered, for it was the girl's
privilege to choose, and she had chosen. He cared
nothing for that now, but he felt it indeed a reproach
to tamely let this man take his horse when he had
all the mountain at his back. There was a sharp
humiliation in his position. He felt the pressure of
public opinion.

"Dad-burn him !" he exclaimed. "Ef I kin make
out ter git a glimge o' him, I'll shoot him dead—
dead !"

He leaned the rifle against the rock. It struck
upon a ledge. A metallic vibration rang out. Again

and again the sound was repeated — now loud, still clanging; now faint, but clear; now soft and away to a doubtful murmur which he hardly was sure that he heard.   Never before had he known such an echo. And suddenly he recollected that this was the great "Talking Rock," famed beyond the limits of Lonesome.   It had traditions as well as echoes.   He remembered vaguely that beneath this cliff there was said to be a cave which was utilized in the manufacture of saltpetre for gunpowder in the War of 1812.

As he looked down the slope below he thought the snow seemed broken — by footprints, was it? With the expectation of a discovery strong upon him, he crept along a wide ledge of the crag, now and then stumbling and sending an avalanche of snow and ice and stones thundering to the foot of the cliff.   He missed his way more than once.   Then he would turn about, laboriously retracing his steps, and try another level of the ledges.   Suddenly before him was the dark opening he sought.   No creature had lately been here.   It was filled with growing bushes and dead leaves and brambles. Looking again down upon the slope beneath, he felt very sure that he saw footprints.

"The old folks useter 'low ez thar war two openings ter this hyar cave," he said.   "Tobe Gryce mought hev hid hyar through a opening down yander on the slope.   But *I'll* go the way ez I hev hearn tell on, an' peek in, an' ef I kin git a glimpse o' him, I'll make him tell me whar that thar filly air, or I'll let daylight through him, sure!" ·

He paused only to bend aside the brambles, then he crept in and took his way along a low, narrow passage. It had many windings, but was without intersections or intricacy. He heard his own steps echoed like a pursuing footfall. His labored breathing returned in sighs from the inanimate rocks. It was an uncanny place, with strange, sepulchral, solemn effects. He shivered with the cold. A draught stole in from some secret crevice known only to the wild mountain winds. The torch flared, crouched before the gust, flared again, then darkness. He hesitated, took one step forward, and suddenly—a miracle!

A soft aureola with gleaming radiations, a low, shadowy chamber, a beast feeding from a manger, and within it a child's golden head.

His heart gave a great throb. Somehow he was smitten to his knees. Christmas Eve! He remembered the day with a rush of emotion. He stared again at the vouchsafed vision. He rubbed his eyes. It had changed.

Only hallucination caused by an abrupt transition from darkness to light; only the most mundane facts of the old troughs and ash-hoppers, relics of the industry that had served the hideous carnage of battle ; only the yellow head of the ranger's brat, who had climbed into one of them, from which the mare was calmly munching her corn.

Yet this was Christmas Eve. And the Child did lie in a manger.

Perhaps it was well for him that his ignorant faith could accept the illusion as a vision charged with all the benignities of peace on earth, good-will toward men. With a keen thrill in his heart, on his knees he drew the charge from his rifle, and flung it down a rift in the rocks. "Chrismus Eve," he murmured.

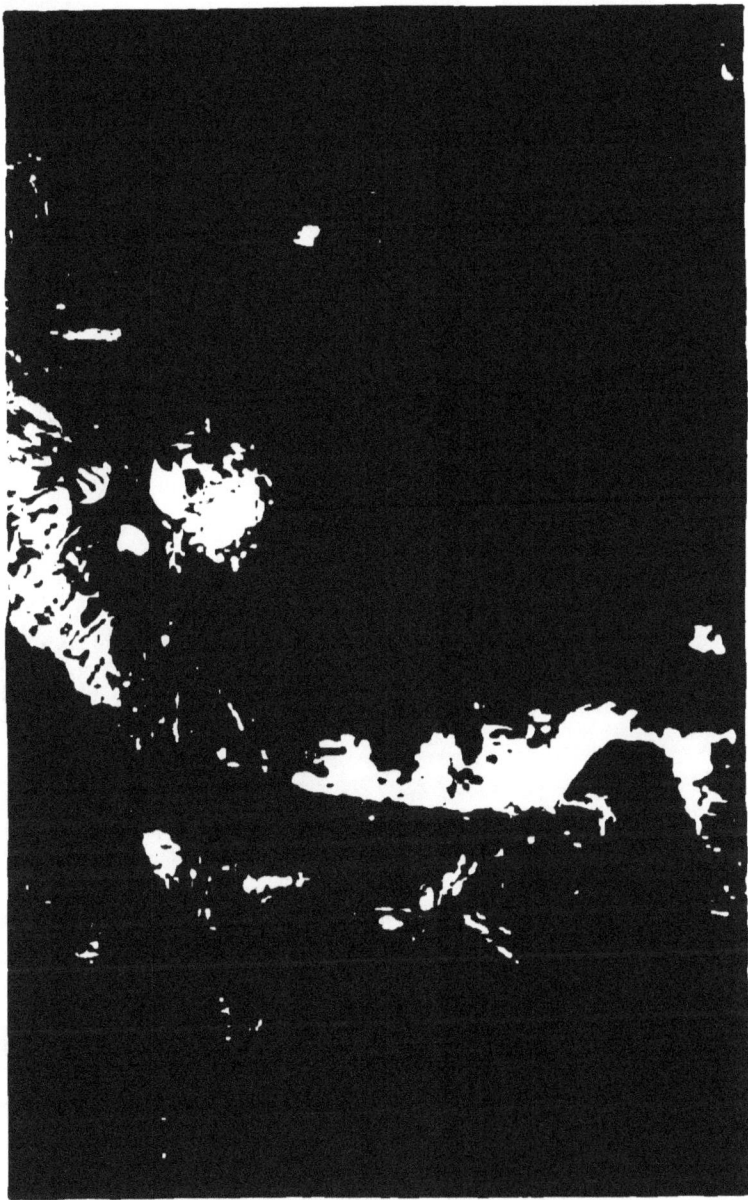

He leaned his empty weapon against the wall, and strode out to the little girl who was perched up on the trough.

"Chrismus gift, Cunnel!" he cried, cheerily. "Ter-morrer's Chrismus."

The echoes caught the word. In vibratory jubilance they repeated it. "Chrismus!" rang from the roof, scintillating with calc-spar; "Chrismus!" sounded from the colonnade of stalactites that hung down to meet the uprising stalagmites; "Chrismus!" repeated the walls incrusted with roses that, shut in from the light and the fresh air of heaven, bloomed forever in the stone. Was ever chorus so sweet as this?

It reached Tobe Gryce, who stood at his improvised corn-bin. With a bundle of fodder still in his arms he stepped forward. There beside the little Colonel and the black mare he beheld a man seated upon an inverted half-bushel measure, peacefully lighting his pipe with a bunch of straws which he kindled at the lantern on the ash-hopper.

The ranger's black eyes were wide with wonder at this intrusion, and angrily flashed. He connected it at once with the attack on the stable. The hair on his low forehead rose bristlingly as he frowned. Yet he realized with a quaking heart that he was helpless. He, although the crack shot of the county, would not have fired while the Colonel was within two yards of his mark for the State of Tennessee.

He stood his ground with stolid courage—a target.

Then, with a start of surprise, he perceived that

12

the intruder was unarmed. Twenty feet away his
rifle stood against the wall.

Tobe Gryce was strangely shaken. He experienced a sudden revolt of credulity. This was surely
a dream.

"Ain't that thar Luke Todd? Why air ye a-waitin' thar?" he called out in a husky undertone.

Todd glanced up, and took his pipe from his
mouth; it was now fairly alight.

"Kase it be Chrismus Eve, Tobe," he said,
gravely.

The ranger stared for a moment; then came
forward and gave the fodder to the mare, pausing
now and then and looking with oblique distrust
down upon Luke Todd as he smoked his pipe.

"I want ter tell ye, Tobe, ez some o' the mounting boys air a-sarchin fur ye outside."

"Who air they?" asked the ranger, calmly.

His tone was so natural, his manner so unsuspecting, that a new doubt began to stir in Luke Todd's
mind.

"What ails ye ter keep the mare down hyar,
Tobe?" he asked, suddenly. "'Pears like ter me
ez that be powerful comical."

"Kase," said Tobe, reasonably, "some durned
horse-thieves kem arter her one night. I fired at
'em. I hain't hearn on 'em sence. An' so I jes
hid the mare."

Todd was puzzled. He shifted his pipe in his
mouth. Finally he said: "Some folks 'lowed ez
ye hed no right ter take up that mare, bein' ez ye
war the ranger."

Tobe Gryce whirled round abruptly. "What

war I a-goin' ter do, then ? Feed the critter fur
nuthin till the triflin' scamp ez owned her kem arter
her ? I couldn't work her 'thout takin' her up an'
hevin her appraised. Thar's a law agin sech. An'
I couldn't git somebody ter toll her off an' take her
up. That ain't fair. What ought I ter hev done ?"

" Wa'al," said Luke, drifting into argument, " the
town-folks 'low ez ye hev got nuthin ter prove it by,
the stray-book an' records bein' burnt. The town-
folks 'low ez ye can't prove by writin' an' sech ez ye
ever tried ter find the owner."

"The town-folks air fairly sodden in foolishness,"
exclaimed the ranger, indignantly.

He drew from his ample pocket a roll of ragged
newspapers, and pointed with his great thumb at a
paragraph. And Luke Todd read by the light of
the lantern the advertisement and description of
the estray printed according to law in the nearest
newspaper.

The newspaper was so infrequent a factor in the
lives of the mountain gossips that this refutation
of their theory had never occurred to them.

The sheet was trembling in Luke Todd's hand ;
his eyes filled. The cavern with its black distances,
its walls close at hand sparkling with delicate points
of whitest light ; the yellow flare of the lantern ; the
grotesque shadows on the ground ; the fair little
girl with her golden hair ; the sleek black mare ;
the burly figure of the ranger—all the scene swayed
before him. He remembered the gracious vision
that had saluted him ; he shuddered at the crime
from which he was rescued. Pity him because he
knew naught of the science of optics ; of the be-

wildering effects of a sudden burst of light upon
the delicate mechanism of the eye; of the vagaries
of illusion.

"Tobe," he said, in a solemn voice — all the
echoes were bated to awed whispers—" I hev been
gin ter view a vision this night, bein' 'twar Chrismus Eve. An' now I want ter shake hands on it
fur peace."

Then he told the whole story, regardless of the
ranger's demonstrations, albeit they were sometimes
violent enough. Tobe sprang up with a snort of
rage, his eyes flashing, his thick tongue stumbling
with the curses crowding upon it, when he realized
the suspicions rife against him at the county town.
But he stood with his clinched hand slowly relaxing,
and with the vague expression which one wears
who looks into the past, as he listened to the recital
of Eugenia's pilgrimage in the snowy wintry dawn.
"Mighty few folks hev got a wife ez set store by
'em like that," Luke remarked, impersonally.

The ranger's rejoinder seemed irrelevant.

"'Genie be a-goin' ter see a powerful differ arter
this," he said, and fell to musing.

Snow, fatigue, and futility destroyed the ardor of
the lynching party after a time, and they dispersed
to their homes. Little was said of this expedition
afterward, and it became quite impossible to find
a man who would admit having joined it. For
the story went the rounds of the mountain that
there had been a mistake as to unfair dealing on
the part of the ranger, and Luke Todd was quite
content to accept from the county treasury half the
sum of the mare's appraisement—with the deduc-

tion of the stipulated per cent.—which Tobe Gryce
had paid, the receipt for which he produced.

The gossips complained, however, that after all
this was settled according to law, Tobe wouldn't
keep the mare, and insisted that Luke should return to him the money he had paid into the treasury, half her value, "bein' so brigaty he wouldn't
own Luke Todd's beast. An' Luke agreed ter so
do; but he didn't want ter be outdone, so fur the
keep o' the filly he gin the Cunnel a heifer. An'
Tobe war mighty nigh tickled ter death fur the
Cunnel ter hev a cow o' her own."

And now when December skies darken above
Lonesome Cove, and the snow in dizzying whirls
sifts softly down, and the gaunt brown leafless
heights are clothed with white as with a garment,
and the wind whistles and shouts shrilly, and above
the great crag loom the distant mountains, and below are glimpsed the long stretches of the valley,
the two men remember the vision that illumined
the cavernous solitudes that night, and bless the
gracious power that sent salvation 'way down to
Lonesome Cove, and cherish peace and good-will
for the sake of a little Child that lay in a manger.

THE
MOONSHINERS AT HOHO-HEBEE FALLS

THE
MOONSHINERS AT HOHO-HEBEE FALLS

I

IF the mission of the little school-house in Holly
Cove was to impress upon the youthful mind a
comprehension and appreciation of the eternal
verities of nature, its site could hardly have been
better chosen. All along the eastern horizon de-
ployed the endless files of the Great Smoky Mountains—blue and sunlit, with now and again the apparition of an unfamiliar peak, hovering like a
straggler in the far-distant rear, and made visible
for the nonce by some exceptional clarification of
the atmosphere; or lowering, gray, stern; or with
ranks of clouds hanging on their flanks, while all
the artillery of heaven whirled about them, and the
whole world quaked beneath the flash and roar of
its volleys. The seasons successively painted the
great landscape—spring, with its timorous touch,
its illumined haze, its tender, tentative green and
gray and yellow; summer, with its flush of comple-
tion, its deep, luscious, definite verdure, and the
golden richness of fruition; autumn, with a full
brush and all chromatic splendors; winter, in melancholy sepia tones, black and brown and many

sad variations of the pallors of white. So high was the little structure on the side of a transverse ridge that it commanded a vast field of sky above the wooded ranges; and in the immediate foreground, down between the slopes which were cleft to the heart, was the river, resplendent with the reflected moods of the heavens. In this deep gorge the winds and the pines chanted like a Greek chorus; the waves continuously murmured an intricate rune, as if conning it by frequent repetition; a bird would call out from the upper air some joyous apothegm in a language which no creature of the earth has learned enough of happiness to translate.

But the precepts which prevailed in the little school - house were to the effect that rivers, except as they flowed as they listed to confusing points of the compass, rising among names difficult to remember, and emptying into the least anticipated body of water, were chiefly to be avoided for their proclivity to drown small boys intent on swimming or angling. Mountains, aside from the desirability of their recognition as forming one of the divisions of land somewhat easily distinguishable by the more erudite youth from plains, valleys, and capes, were full of crags and chasms, rattlesnakes and vegetable poisons, and a further familiarity with them was liable to result in the total loss of the adventurous—to see friends, family, and home no more.

These dicta, promulgated from the professorial chair, served to keep the small body of callow humanity, with whose instruction Abner Sage was

intrusted by the State, well within call and out of
harm's way during the short recesses, while under
his guidance they toddled along the rough road
that leads up the steeps to knowledge.   But one
there was who either bore a charmed life or possessed an unequalled craft in successfully defying
danger; who fished and swam with impunity; who
was ragged and torn from much climbing of crags;
whose freckled face bore frequent red tokens of an
indiscriminate sampling of berries.   It is too much
to say that Abner Sage would have been glad to
have his warnings made terrible by some bodily
disaster to the juvenile dare-devil of the school,
but Leander Yerby's disobedient incredulity as to
the terrors that menaced him, and his triumphant
immunity, fostered a certain grudge against him.
Covert though it was, unrecognized even by Sage
himself, it was very definitely apparent to Tyler
Sudley when sometimes, often, indeed, on his way
home from hunting, he would pause at the school-
house window, pulling open the shutter from the
outside, and gravely watch his protégé, who stood
spelling at the head of the class.

For Leander Yerby's exploits were not altogether
those of a physical prowess.   He was a mighty
wrestler with the multiplication table.   He had
met and overthrown the nine line in single-hand-
ed combat.   He had attained unto some interesting knowledge of the earth on which he lived,
and could fluently bound countries with neatness
and precision, and was on terms of intimacy with
sundry seas, volcanoes, islands, and other sizable
objects.   The glib certainty of his contemptuous

familiarity with the alphabet and its untoward com-
binations, as he flung off words in four syllables in
his impudent chirping treble, seemed something
uncanny, almost appalling, to Tyler Sudley, who
could not have done the like to save his stalwart
life. He would stare dumfounded at the erudite
personage at the head of the class; Leander's bare
feet were always carefully adjusted to a crack between the puncheons of the floor, literally "toeing
the mark"; his broad trousers, frayed out liberally
at the hem, revealed his skinny and scarred little
ankles, for his out-door adventures were not without a record upon the more impressionable portions
of his anatomy; his waistband was drawn high up
under his shoulder-blades and his ribs, and girt
over the shoulders of his unbleached cotton shirt
by braces, which all his learning did not prevent
him from calling "galluses"; his cut, scratched,
calloused hands were held stiffly down at the side
seams in his nether garments in strict accordance
with the regulations. But rules could not control
the twinkle in his big blue eyes, the mingled effron-
tery and affection on his freckled face as he perceived the on-looking visitor, nor hinder the wink,
the swiftly thrust-out tongue, as swiftly withdrawn,
the egregious display of two rows of dishevelled
jagged squirrel teeth, when once more, with an off-
hand toss of his tangled brown hair, he nimbly
spelled a long twisted-tailed word, and leered capably at the grave intent face framed in the window.

"Why, Abner!" Tyler Sudley would break out,
addressing the teacher, all unmindful of scholastic
etiquette, a flush of pleasure rising to his swarthy

cheek as he thrust back his wide black hat on his long dark hair and turned his candid gray eyes, all aglow, upon the cadaverous, ascetic preceptor, "ain't Lee-yander a-gittin' on powerful, *powerful* fas' with his book?"

"Not in enny ways so special," Sage would reply in cavalier discouragement, his disaffected gaze resting upon the champion scholar, who stood elat·ed, confident, needing no commendation to assure him of his pre-eminence ; "but he air disobejient, an' turr'ble, turr'ble bad."

The nonchalance with which Leander Yerby hearkened to this criticism intimated a persuasion that there were many obedient people in this world, but few who could so disport themselves in the intricacies of the English language ; and Sudley, as he plodded homeward with his rifle on his shoulder, his dog running on in advance, and Leander pattering along behind, was often moved to add the weight of his admonition to the teacher's reproof.

"Lee - yander," he would gently drawl, "ye mustn't be so bad, honey ; ye *mustn't* be so turr'ble bad."

"Naw, ma'am, I won't," Leander would cheerily pipe out, and so the procession would wend its way along.

For he still confused the gender in titles of re-spect, and from force of habit he continued to do so in addressing Tyler Sudley for many a year after he had learned better.

These lapses were pathetic rather than ridiculous in the hunter's ears. It was he who had taught Leander every observance of verbal humility tow-

ard his wife, in the forlorn hope of propitiating her
in the interest of the child, who, however, with his
quick understanding that the words sought to do
honor and express respect, had of his own accord
transferred them to his one true friend in the household. The only friend he had in the world, Sudley
often felt, with a sigh over the happy child's forlorn
estate. And, with the morbid sensitiveness peculiar to a tender conscience, he winced under the
knowledge that it was he who, through wrong-
headedness or wrongheartedness, had contrived to
make all the world besides the boy's enemy. Both
wrongheaded and wronghearted he was, he sometimes told himself. For even now it still seemed
to him that he had not judged amiss, that only the
perversity of fate had thwarted him. Was it so fan-
tastically improbable, so hopeless a solace that he
had planned, that he should have thought his wife
might take comfort for the death of their own child
in making for its sake a home for another, orphaned,
forlorn, a burden, and a glad riddance to those
into whose grudging charge it had been thrown?
This bounty of hope and affection and comfort had
seemed to him a free gift from the dead baby's
hands, who had no need of it since coming into its
infinite heritage of immortality, to the living waif,
to whom it was like life itself, since it held all the
essential values of existence. The idea smote him
like an inspiration. He had ridden twenty miles
in a snowy night to beg the unwelcome mite from
the custody of its father's half-brothers, who were
on the eve of moving to a neighboring county with
all their kin and belongings.

Tyler Sudley was a slow man, and tenacious of impressions. He could remember every detail of the events as they had happened—the palpable surprise, the moment of hesitation, the feint of denial which successively ensued on his arrival. It mattered not what the season or the hour—he could behold at will the wintry dawn, the deserted cabin, the glow of embers dying on the hearth within; the white-covered wagon slowly a-creak along the frozen road beneath the gaunt, bare, overhanging trees, the pots and pans as they swung at the rear, the bucket for water swaying beneath, the mounted men beside it, the few head of swine and cattle driven before them. Years had passed, but he could feel anew the vague stir of the living bundle which he held on the pommel of his saddle, the sudden twist it gave to bring its inquiring, apprehensive eyes, so large in its thin, lank-jawed, piteous little countenance, to bear on his face, as if it understood its transfer of custody, and trembled lest a worse thing befall it. One of the women stopped the wagon and ran back to pin about its neck an additional wrapping, an old red-flannel petticoat, lest it should suffer in its long, cold ride. His heart glowed with vicarious gratitude for her forethought, and he shook her hand warmly and wished her well, and hoped that she might prosper in her new home, and stood still to watch the white wagon out of sight in the avenue of the snow-laden trees, above which the moon was visible, a-journeying too, swinging down the western sky.

Laurelia Sudley sat in stunned amazement when, half-frozen, but triumphant and flushed and full

of his story, he burst into the warm home atmosphere, and put the animated bundle down upon the
hearth-stone in front of the glowing fire.   For one
moment she met its forlorn gaze out of its peaked
and pinched little face with a vague hesitation in
her own worn, tremulous, sorrow - stricken eyes.
Then she burst into a tumult of tears, upbraiding
her husband that he could think that another child
could take the place of her dead child — all the
dearer because it was dead ; that she could play
the traitor to its memory and forget her sacred
grief ; that she could do aught as long as she
should live but sit her down to bewail her loss,
every tear a tribute, every pang its inalienable
right, her whole smitten existence a testimony to
her love.   It was in vain that he expostulated.   The
idea of substitution had never entered his mind.
But he was ignorant, and clumsy of speech, and unaccustomed to analyze his motives.   He could not
put into words his feeling that to do for the wel-
fare of this orphaned and unwelcome little creature
all that they would have done for their own was in
some sort a memorial to him, and brought them
nearer to him—that she might find in it a satisfaction, an occupation—that it might serve to fill her
empty life, her empty arms.

But no ! She thought, and the neighbors thought,
and after a time Tyler Sudley came to think also,
that he had failed in the essential duty to the dead
—that of affectionate remembrance ; that he was
recreant, strangely callous.   They all said that he
had seemed to esteem one baby as good as another,
and that he was surprised that his wife was not

consoled for the loss of her own child because he
took it into his head to go and toll off the Yerby
baby from his father's half-brothers "ez war movin'
away an' war glad enough ter get rid o' one head o'
human stock ter kerry, though, *bein' human*, they
oughter been ashamed ter gin him away like a
puppy-dog, or an extry cat, all hands consarned."

From the standpoint she had taken Laurelia had
never wavered. It was an added and a continual
reproach to her husband that all the labor and
care of the ill-advised acquisition fell to her share.
She it was who must feed and clothe and tend the
gaunt little usurper; he needs must be accorded
all the infantile prerogatives, and he exacted much
time and attention. Despite the grudging spirit in
which her care was given she failed in no essential,
and presently the interloper was no longer gaunt
or pallid or apprehensive, but grew pink and che-
rubic of build, and arrogant of mind. He had no
sensitive sub-current of suspicion as to his welcome ;
he filled the house with his gay babbling, and if no
maternal chirpings encouraged the development of
his ideas and his powers of speech, his cheerful
spirits seemed strong enough to thrive on their
own stalwart endowments. His hair began to curl,
and a neighbor, remarking on it to Laurelia, and
forgetting for the moment his parentage, said, in
admiring glee, twining the soft tendrils over her finger, that Mrs. Sudley had never before had a child so
well-favored as this one. From this time forth was
infused a certain rancor into his foster-mother's
spirit toward him. Her sense of martyrdom was
complete when another infant was born and died,

13

leaving her bereaved once more to watch this stranger grow up in her house, strong and hearty, and handsomer than any child of hers had been.

The mountain gossips had their own estimate of her attitude.

"I ain't denyin' but what she hed nat'ral feelin' fur her own chil'ren, bein' dead," said the dame who had made the unfortunate remark about the curling hair, "but Laurelia Sudley war always a contrary-minded, lackadaisical kind o' gal afore she war married, sorter set in opposition, an' now ez she ain't purty like she useter was, through cryin' her eyes out, an' gittin' sallow-complected an' bony, I kin notice her contrariousness more. Ef Tyler hedn't brung that chile home, like ez not she'd hev sot her heart on borryin' one herself from somebody. Lee-yander ain't in nowise abused, ez I kin see — ain't acquainted with the rod, like the Bible say he oughter be, an' ennybody kin see ez Laurelia don't like the name he gin her, yit she puts up with it. She larnt him ter call Ty 'Cap'n,' bein' she's sorter proud of it, 'kase Ty war a cap'n of a critter company in the war: 'twarn't sech a mighty matter nohow; he jes got ter be cap'n through the other off'cers bein' killed off. An' the leetle boy got it twisted somehows, an' calls *her* 'Cap'n' an' Ty 'Neighbor,' from hearin' old man Jeemes, ez comes in constant, givin' him that old-fashioned name. 'Cap'n' 'bout fits Laurelia, though, an' that's a fac'."

Laurelia's melancholy ascendency in the household was very complete. It was characterized by no turbulence, no rages, no long-drawn argument or objurgation; it expressed itself only in a settled spirit of disaffection, a pervasive suggestion of martyrdom, silence or sighs, or sometimes a depressing singing of hymn tunes. For her husband had long ago ceased to remonstrate, or to seek to justify himself. It was with a spirit of making amends that he hastened to concede every point of question, to defer to her preference in all matters, and Laurelia's sway grew more and more absolute as the years wore on. Leander Yerby could remember no other surroundings than the ascetic atmosphere of his home. It had done naught apparently to quell the innate cheerfulness of his spirit. He evidently took note, however, of the different standpoint of the "Captain" and his "Neighbor," for although he was instant in the little manifestations of respect toward her which he had been taught, his childish craft could not conceal their spuriousness.

"That thar boy treats me ez ef I war a plumb idjit," Laurelia said one day, moved to her infrequent anger. "Tells me, 'Yes, ma'am, cap'n,' an' 'Naw, ma'am, cap'n,' jes ter quiet me—like folks useter do ter old Ed'ard Green, ez war in his dotage—an' then goes along an' does the very thing I tell him not ter do."

Sudley looked up as he sat smoking his pipe by the fire, a shade of constraint in his manner, and a contraction of anxiety in his slow, dark eyes, never quite absent when she spoke to him aside of Leander.

She paused, setting her gaunt arms akimbo, and wearing the manner of one whose kindly patience is beyond limit abused. "Kems in hyar, he do, a-totin' a fiddle. An' I says, 'Lee-yander Yerby, don't ye know that thar thing's the devil's snare?' 'Naw, ma'am, cap'n,' he says, grinnin' like a imp; 'it's *my* snare, fur I hev bought it from Peter Teazely fur two rabbits what I cotch in my trap, an' my big red rooster, an' a bag o' seed pop-corn, an' the only hat I hev got in the worl'.' An' with that the consarn gin sech a yawp, it plumb went through my haid. An' then the critter jes tuk ter a-bowin' it back an' forth, a-playin' 'The Chicken in the Bread-trough' like demented, a-dancin' off on fust one foot an' then on t'other till the puncheons shuck. An' I druv him out the house. I won't stan' none o' Satan's devices hyar! I tole him he couldn't fetch that fiddle hyar whenst he kems home ter-night, an' I be a-goin' ter make him a sun-bonnet or a nightcap ter wear stiddier his hat that he traded off."

She paused.

Her husband had risen, the glow of his pipe fading in his unheeding hand, his excited eyes fixed upon her. "Laurely," he exclaimed, "ye ain't meanin' ez that thar leetle critter could play a chune fust off on a fiddle 'thout no larnin'!"

She nodded her head in reluctant admission.

He opened his mouth once or twice, emitting no sound. She saw how his elation, his spirit of commendation, his pride, set at naught her displeasure, albeit in self-defence, perchance, he dared not say a word. With an eye alight and an absorbed face, he laid his pipe on the mantel-piece, and silently took his way out of the house in search of the youthful musician.

Easily found! The racked and tortured echoes were all aquake within half a mile of the spot where, bareheaded, heedless of the threatened ignominy alike of sun-bonnet or nightcap, Leander sat in the flickering sunshine and shadow upon a rock beside the spring, and blissfully experimented with all the capacities of catgut to produce sound.

"Listen, Neighbor!" he cried out, descrying Tyler Sudley, who, indeed, could do naught else—"*listen!* Ye won't hear much better fiddlin' this side o' kingdom come!" And with glad assurance he capered up and down, the bow elongating the sound to a cadence of frenzied glee, as his arms sought to accommodate the nimbler motions of his legs.

Thus it was the mountaineers later said that Leander fell into bad company. For, the fiddle being forbidden in the sober Laurelia's house, he must needs go elsewhere to show his gift and his growing skill, and he found a welcome fast enough. Before he had advanced beyond his stripling youth, his untutored facility had gained a rude mastery over the instrument; he played with a sort of fascination and spontaneity that endeared his art to his uncritical audiences, and his endowment was held as something wonderful. And now it was that Laurelia, hearing him, far away in the open air, play once a plaintive, melodic strain, fugue-like with the elfin echoes, felt a strange soothing in the sound, found tears in her eyes, not all of pain but of sad pleasure, and assumed thenceforth something of the port of a connoisseur. She said she "couldn't abide a fiddle jes sawed helter-skelter by them ez hedn't larned, but ter play saaft an' slow an' solemn, and no dancin' chune, no frolic song—she warn't set agin that at all." And she desired of Leander a repetition of this sunset motive that evening when he had come home late, and she discovered him hiding the obnoxious instrument under the porch. But in vain. He did not remember it. It was some vague impulse, as unconsciously voiced as the dreaming bird's song in the sudden half-awake intervals of the night. Over and again, as he stood by the porch, the violin in his arms, he touched the strings tentatively, as if, perchance, being so alive, they might of their own motion recall the strain that had so lately thrilled along them.

He had grown tall and slender. He wore boots to his knees now, and pridefully carried a "shootin'-iron" in one of the long legs—to his great discomfort. The freckles of his early days were merged into the warm uniform tint of his tanned complexion. His brown hair still curled; his shirt-collar fell away from his throat, round and full and white—the singer's throat—as he threw his head backward and cast his large roving eyes searchingly along the sky, as if the missing strain had wings.

The inspiration returned no more, and Laurelia experienced a sense of loss. "Some time, Leeyander, ef ye war ter kem acrost that chune agin, try ter set it in yer remembrance, an' play it whenst ye kem home," she said, wistfully, at last, as if this errant melody were afloat somewhere in the vague realms of sound, where one native to those haunts might hope to encounter it anew.

"Yes, ma'am, cap'n, I will," he said, with his

facile assent. But his tone expressed slight intention, and his indifference bespoke a too great wealth
of "chunes"; he could feel no lack in some unre-
membered combination, sport of the moment, when
another strain would come at will, as sweet perchance, and new.

She winced as from undeserved reproach when
presently Leander's proclivities for the society of
the gay young blades about the countryside, sometimes reputed "evil men," were attributed to this
exile of the violin from the hearth-stone. She
roused herself to disputation, to indignant repudia-
tion.

"They talk ez ef it war *me* ez led the drinkin',
an' the gamin', an' the dancin', and sech, ez goes
on in the Cove, 'kase whenst Lee-yander war about
fryin' size I wouldn't abide ter hev him a-sawin' away
on the fiddle in the house enough ter make me deef
fur life. At fust the racket of it even skeered Towse
so he wouldn't come out from under the house fur
two days an' better; he jes sot under thar an'
growled, an' shivered, an' showed his teeth ef ennybody spoke ter him. Nobody don't like Lee-yander's performin' better'n I do whenst he plays them
saaft, slippin'-away, slow medjures, ez sound plumb
religious—ef 'twarn't a sin ter say so. Naw, sir,
ef ennybody hev sot Lee-yander on ter evil ways
'twarn't me. My conscience be clear."

Nevertheless she was grievously ill at ease when
one day there rode up to the fence a tall, gaunt, ill-
favored man, whose long, lean, sallow countenance,
of a Pharisaic cast, was vaguely familiar to her, as
one recognizes real lineaments in the contortions of

a caricature or the bewilderments of a dream. She felt as if in some long-previous existence she had seen this man as he dismounted at the gate and came up the path with his saddle-bags over his arm. But it was not until he mustered an unready, unwilling smile, that had of good-will and geniality so slight an intimation that it was like a spasmodic grimace, did she perceive how time had deepened tendencies to traits, how the inmost thought and the secret sentiment had been chiselled into the face in the betrayals of the sculpture of fifteen years.

"Nehemiah Yerby!" she exclaimed. "I would hev knowed ye in the happy land o' Canaan."

"Let's pray we may all meet thar, Sister Sudley," he responded. "Let's pray that the good time may find none of us unprofitable servants."

Mrs. Sudley experienced a sudden recoil. Not that she did not echo his wish, but somehow his manner savored of an exclusive arrogation of piety and a suggestion of reproach.

"That's my prayer," she retorted, aggressively. "Day an' night, that's my prayer."

"Yes'm, fur us an' our households, Sister Sudley —we mus' think o' them c'mitted ter our charge."

She strove to fling off the sense of guilt that oppressed her, the mental attitude of arraignment. He was a young man when he journeyed away in that snowy dawn. She did not know what changes had come in his experience. Perchance his effervescent piety was only a habit of speech, and had no significance as far as she was concerned. The suspicion, however, tamed her in some sort. She attempted no retort. With a mechanical, reluctant smile, ill adjusted to her sorrow-lined face, she made an effort to assume that the greeting had been but the conventional phrasings of the day.

"Kem in, kem in, Nehemiah; Tyler will be glad ter see ye, an' I reckon ye will be powerful interested ter view how Lee-yander hev growed an' prospered."

She felt as if she were in some terrible dream as she beheld him slowly wag his head from side to side. He had followed her into the large main room of the cabin, and had laid his saddle-bags down by the side of the chair in which he had seated himself, his elbows on his knees, his hands held out to the flickering blaze in the deep chimney-place, his eyes significantly narrowing as he gazed upon it.

"Naw, Sister Sudley," he wagged his head more mournfully still. "I kin but grieve ter hear how my nevy Lee-yander hev 'prospered,' ez ye call it, an' I be s'prised ye should gin it such a name. Oh-h-h, Sister Sudley!" in prolonged and dreary vocative, "I 'lowed ye war a godly woman. I knowed yer name 'mongst the church-goers an' the church-members." A faint flush sprang into her delicate faded cheek; a halo encircled this repute of sanctity; she felt with quivering premonition that it was about to be urged as a testimony against her. "Elsewise I wouldn't hev gin my cornsent ter hev lef' the leetle lam', Lee-yander, in yer fold. Precious, precious leetle lam'!"

Poor Laurelia! Were it not that she had a sense of fault under the scathing arraignment of her mo-

tives, her work, and its result, although she scarcely saw how she was to blame, that she had equally
with him esteemed Leander's standpoint iniquitous,
she might have made a better fight in her own interest.   Why she did not renounce the true culprit
as one on whom all godly teachings were wasted,
and, adopting the indisputable vantage-ground of
heredity, carry the war into the enemy's country,
ascribing Leander's shortcomings to his Yerby
blood, and with stern and superior joy proclaiming
that he was neither kith nor kin of hers, she wondered afterward, for this valid ground of defence
did not occur to her then.   In these long mourning years she had grown dull; her mental processes were either a sad introspection or reminiscence.   Now she could only take into account her
sacrifices of feeling, of time, of care; the illnesses
she had nursed, the garments that she had made
and mended — ah, how many! laid votive on the
altar of Leander's vigor and his agility, for as he
scrambled about the crags he seemed, she was
wont to say, to climb straight out of them.   The
recollection of all this—the lesser and unspiritual
maternal values,. perchance, but essential—surged
over her with bitterness; she lost her poise, and
fell a-bickering.

"'Precious leetle lam','" she repeated, scornfully.
"Precious he mus' hev been!   Fur when ye lef'
him he hedn't a whole gyarmint ter his back, an'
none but them that kivered him."

Nehemiah Yerby changed color slightly as the
taunt struck home, but he was skilled in the more
æsthetic methods of argument.

"We war pore — mighty pore indeed, Sister Sudley."

Now, consciously in the wrong, Sister Sudley, with true feminine inconsistency, felt better. She retorted with bravado.

"Needle an' thread ain't 'spensive nowhar ez I knows on, an' the gov'mint hev sot no tax on saaft home-made soap, so far ez hearn from."

She briskly placed her chair, a rude rocker, the seat formed of a taut-stretched piece of ox-hide, beside the fire, and took up her knitting. A sock for Leander it was—one of many of all sizes. She remembered the first that she had measured for the bare pink toes which he had brought there, forlorn candidates for the comfortable integuments in which they were presently encased, and how she had morbidly felt that every stitch she took was a renunciation of her own children, since a stranger was honored in their place. The tears came into her eyes. It was only this afternoon that she had experienced a pang of self-reproach to realize how near happiness she was—as near as her temperament could approach. But somehow the air was so soft; she could see from where she sat how the white velvet buds of the aspen-trees in the dooryard had lengthened into long, cream-tinted, furry tassels; the maples on the mountain-side lifted their red flowering boughs against the delicate blue sky; the grass was so green; the golden candlesticks bunched along the margin of the path to the rickety gate were all a-blossoming. The sweet appeal of spring had never been more insistent, more coercive. Somehow peace, and a placid content, seemed as essential incidents in the inner life as the growth of the grass anew, the bursting of the bud, or the soft awakening of the zephyr. Even within the house, the languors of the fire drowsing on the hearth, the broad bar of sunshine across the puncheon floor, so slowly creeping away, the sense of the vernal lengthening of the pensive afternoon, the ever-flitting shadow of the wren building under the eaves, and its iterative gladsome song breaking the fireside stillness, partook of the serene beatitude of the season and the hour. The visitor's drawling voice rose again, and she was not now constrained to reproach herself that she was too happy.

"Yes'm, pore though we war then — an' we couldn't look forward ter the Lord's prosperin' us some sence—we never would hev lef' the precious leetle lam'"—his voice dwelt with unvanquished emphasis upon the obnoxious words — "'mongst enny but them persumed ter be godly folks. Tyler war a toler'ble good soldier in the war, an' hed a good name in the church, but *ye* war persumed to be a plumb special Christian with no pledjure in this worl'."

Laurelia winced anew. This repute of special sanctity was the pride of her ascetic soul. Few of the graces of life or of the spirit had she coveted, but her pre-eminence as a religionist she had fostered and cherished, and now through her own deeds of charity it seemed about to be wrested from her.

"Lee-yander Yerby hev larnt nuthin' but good in this house, an' all my neighbors will tell you the same word. The Cove 'lows I hev been *too* strict."

Nehemiah was glancing composedly about the room. "That thar 'pears ter be a fiddle on the wall, ain't it, Mis' Sudley?" he said, with an incidental air and the manner of changing the subject.

Alack, for the æsthetic perversion! Since the playing of those melancholy minor strains in that red sunset so long ago, which had touched so responsive a chord in Laurelia's grief-worn heart, the crazy old fiddle had been naturalized, as it were, and had exchanged its domicile under the porch for a position on the wall. It was boldly visible, and apparently no more ashamed of itself than was the big earthen jar half full of cream, which was placed close to the fireplace on the hearth in the hope that its contents might become sour enough by to-morrow to be churned.

Laurelia looked up with a start at the instrument, red and lustrous against the brown log wall, its bow poised jauntily above it, and some glistening yellow reflection from the sun on the floor playing among the strings, elusive, soundless fantasies.

Her lower jaw dropped. She was driven to her last defences, and sore beset. "It air a fiddle," she said, slowly, at last, and with an air of conscientious admission, as if she had had half a mind to deny it. "A fiddle the thing air." Then, as she collected her thoughts, "Brother Pete Vickers 'lows ez he sees no special sin in playin' the fiddle. He 'lows ez in some kentries—I disremember whar —they plays on 'em in church, quirin' an' hymn chunes an' sech."

Her voice faltered a little; she had never thought

to quote this fantasy in her own defence, for she secretly believed that old man Vickers must have been
humbugged by some worldly brother skilled in drawing the long bow himself.

Nehemiah Yerby seemed specially endowed with
a conscience for the guidance of other people, so
quick was he to descry and pounce upon their shortcomings. If one's sins are sure to find one out, there
is little doubt but that Brother Nehemiah would be
on the ground first.

"Air 'you-uns a-settin' under the preachin' o'
Brother Peter Vickers?" he demanded in a sepulchral voice.

"Naw, naw," she was glad to reply. "'Twar
onderstood ez Brother Vickers wanted a call ter the
church in the Cove, bein' ez his relations live hyar-
abouts, an' he kem up an' preached a time or two.
But he didn't git no call. The brethren 'lowed
Brother Vickers war too slack in his idees o' religion. Some said his hell warn't half hot enough.
Thar air some powerful sinners in the Cove, an'
nuthin' but good live coals an' a liquid blazin' fire
air a-goin' ter deter them from the evil o' thar
ways. So Brother Vickers went back the road he
kem."

She knit off her needle while, with his head still
bent forward, Nehemiah Yerby sourly eyed her, feeling himself a loser with Brother Vickers, in that he
did not have the reverend man's incumbency as a
grievance.

"He 'pears ter me ter see mo' pleasure in religion
'n penance, ennyhow," he observed, bitterly. "An'
the Lord knows the bes' of us air sinners."

"An' he laughs loud an' frequent—mightily like a sinner," she agreed. "An' whenst he prays, he prays loud an' hearty, like he jes expected ter git what he axed fur sure's shootin'. Some o' the bretherin' sorter taxed him with his sperits, an' he 'lowed he couldn't holp but be cheerful whenst he hed the Lord's word fur it ez all things work tergether fur good. An' he laffed same ez ef they hedn't spoke ter him serious."

"Look at that, now!" exclaimed Nehemiah. "An' that thar man ez good ez dead with the heart-disease."

Laurelia's eyes were suddenly arrested by his keen, pinched, lined face. What there was in it to admonish her she could hardly have said, nor how it served to tutor her innocent craft.

"I ain't so sure 'bout Brother Vickers bein' so wrong," she said, slowly. "He 'lowed ter me ez I hed spent too much o' my life a-sorrowin', 'stiddier a - praisin' the Lord for his mercies." Her face twitched suddenly; she could not yet look upon her bereavements as mercies. "He 'lowed I would hev been a happier an' a better 'oman ef I hed took the evil ez good from the Lord's hand, fur in his sendin' it's the same. An' I know that air a true word. An' that's what makes me 'low what he said war true 'bout'n that fiddle; that I ought never ter hev pervented the boy from playin' 'round home an' sech, an' 'twarn't no sin but powerful comfortable an' pleasurable ter set roun' of a cold winter night an' hear him play them slow, sweet, dyin'-away chunes—" She dropped her hands, and gazed with the rapt eyes of remembrance through the window

at the sunset clouds which, gathering red and purple and gold on the mountain's brow, were reflected
roseate and amethyst and amber at the mountain's
base on the steely surface of the river. " Brother
Vickers 'lowed he never hearn sech in all his life.
It brung the tears ter his eyes—it surely did."

" He'd a heap better be weepin' fur them black
sheep o' his congregation an' fur Lee-yander's short-
comin's, fur ez fur ez I kin hear he air about ez black
a sheep ez most pastors want ter wrestle with fur
the turnin' away from thar sins. Yes'm, Sister Sudley, that's jes what p'inted out my jewty plain afore
my eyes, an' I riz up an' kem ter be instant in a-doin' of it. 'I'll not leave my own nevy in the tents
o' sin,' I sez. ' I hev chil'n o' my own, hearty feeders
an' hard on shoe-leather, ter support, but I'll not
grudge my brother's son a home.' Yes, Laurely
Sudley, I hev kem ter kerry him back with me. Yer
jewty ain't been done by him, an' I'll leave him a
dweller in the tents o' sin no longer."

His enthusiasm had carried him too far. Laurelia's face, which at first seemed turning to stone
as she gradually apprehended his meaning and his
mission, changed from motionless white to a tremulous scarlet while he spoke, and when he ceased she
retorted herself as one of the ungodly.

" Ye mus' be mighty ambitious ter kerry away a
skin full o' broken bones ! Jes let Tyler Sudley
hear ez ye called his house the tents o' the ungodly,
an' that ye kem hyar a-faultin' me, an' tellin' me ez
I 'ain't done my jewty ennywhar or ennyhow !" she
exclaimed, with a pride which, as a pious saint, she
had never expected to feel in her husband's reputa-

tion as a high-tempered man and a "mighty handy
fighter," and with implicit reliance upon both endowments in her quarrel.

"Only in a speritchual sense, Sister Sudley," Nehemiah gasped, as he made haste to qualify his as-
severation. "I only charge you with havin' sp'iled
the boy; ye hev sp'iled him through kindness ter
him, an' not *ye* so much ez 'Ty. Ty never hed so
much ez a dog that would mind him! His dog
wouldn't answer call nor whistle 'thout he war so
disposed. *I* never faulted ye, Sister Sudley; 'twar
jes Ty I faulted. I know Ty."

He knew, too, that it was safer to call Ty and his
doings in question, big and formidable and belligerent though he was, than his meek-mannered, melancholy, forlorn, and diminutive wife. Nehemiah
rose up and walked back and forth for a moment
with an excited face and a bent back, and a sort of
rabbit-like action. "Now, I put it to you, Sister
Sudley, air Ty a-makin' that thar boy plough ter-
day?—jes *be-you-ti-ful* field weather!"

Sister Sudley, victorious, having regained her
normal position by one single natural impulse of
self-assertion, not as a religionist, but as Tyler Sudley's wife, and hence entitled to all the show of respect which that fact unaided could command, sat
looking at him with a changed face—a face that
seemed twenty years younger; it had the expression it wore before it had grown pinched and ascetic and insistently sorrowful; one might guess
how she had looked when Tyler Sudley first went
up the mountain "a-courtin'." She sought to assume no other stand-point. Here she was in-

14

trenched. She shook her head in negation. The affair was none of hers. Ty Sudley could take ample care of it.

Nehemiah gave a little skip that might suggest a degree of triumph. "Aha, not ploughin'! But *Ty* is ploughin'. I seen him in the field. An' Lee-yander ain't ploughin'! An' how did I know? Ez I war a-ridin' along through the woods this mornin' I kem acrost a striplin' lad a-walkin' through the undergrowth ez onconsarned ez a killdee an' ez nimble. An' under his chin war a fiddle, an' his head war craned down ter it." He mimicked the attitude as he stood on the hearth. "He never looked up wunst. Away he walked, light ez a plover, an' *a-ping, pang, ping, pang,*" in a high falsetto, "went that fiddle! I war plumb 'shamed fur the critters in the woods ter view sech idle sinfulness, a ole ow*el*, a-blinkin' down out'n a hollow tree, kem ter see what *ping, pang, ping, pang* meant, an' thar war a rabbit settin' up on two legs in the bresh, an' a few stray razor-back hawgs; I tell ye I war mortified 'fore even sech citizens ez them, an' a lazy, impident-lookin' dog ez followed him."

"How did ye know 'twar Lee-yander?" demanded Mrs. Sudley, recognizing the description perfectly, but after judicial methods requiring strict proof.

"Oh-h! by the fambly favor," protested the gaunt and hard-featured Nehemiah, capably. "I knowed the Yerby eye."

"He hev got his mother's eyes." Mrs. Sudley had certainly changed her stand-point with a vengeance. "He hev got his mother's *be-you-ti-ful blue* eyes and her curling, silken brown hair—sorter red; little Yerby in *that*, mebbe; but sech eyes, an' sech lashes, an' sech fine curling hair ez none o' yer fambly ever hed, or ever will."

"Mebbe so. I never seen him more'n a minit. But he might ez well hev a *be-you-ti-ful* curlin' nose, like the elephint in the show, for all the use he air, or I be afeard air ever likely ter be."

Tyler Sudley's face turned gray, despite his belligerent efficiencies, when his wife, hearing the clank of the ox-yoke as it was flung down in the shed outside, divined the home-coming of the ploughman and his team, and slipped out to the barn with her news. She realized, with a strange enlightenment as to her own mental processes, what angry jealousy the look on his face would have roused in her only so short a time ago—jealousy for the sake of her own children, that any loss, any grief, should be poignant and pierce his heart save for them. Now she was sorry for him; she felt with him.

But as he continued silent, and only stared at her dumfounded and piteous, she grew frightened —she knew not of what.

"Shucks, Ty!" she exclaimed, catching him by the sleeve with the impluse to rouse him, to awaken him, as it were, to his own old familiar identity; "ye ain't 'feared o' that thar snaggle-toothed skeer-crow in yander; he would be plumb comical ef he didn't look so mean-natured an' sech a hypercrite."

He gazed at her, his eyes eloquent with pain.

"Laurely!" he gasped, "this hyar thing plumb

knocks me down; it jes takes the breath o' life out'n me!"

She hesitated for a moment. Any anxiety, any trouble, seemed so incongruous with the sweet spring-tide peace in the air, that one did not readily take it home to heart. Hope was in the atmosphere like an essential element; one might call it oxygen or caloric or vitality, according to the tendency of mind and the habit of speech. But the heart knew it, and the pulses beat strongly responsive to it. Faith ruled the world. Some tiny bulbous thing at her feet that had impeded her step caught her attention. It was coming up from the black earth, and the buried darkness, and the chill winter's torpor, with all the impulses of confidence in the light without, and the warmth of the sun, and the fresh showers that were aggregating in the clouds somewhere for its nurture—a blind inanimate thing like that! But Tyler Sudley felt none of it; the blow had fallen upon him, stunning him. He stood silent, looking gropingly into the purple dusk, veined with silver glintings of the moon, as if he sought to view in the future some event which he dreaded, and yet shrank to see.

She had rarely played the consoler, so heavily had she and all her griefs leaned on his supporting arm. It was powerless now. She perceived this, all dismayed at the responsibility that had fallen upon her. She made an effort to rally his courage. She had more faith in it than in her own.

"'Feard o' *him!*" she exclaimed, with a sharp tonic note of satire. "Kem in an' view him."

"Laurely," he quavered, "I oughter hev got it

down in writin' from him ; I oughter made him sign
papers agreein' fur me ter keep the boy till he
growed ter be his own man."

She, too, grew pale. "Ye ain't meanin' ter let
him take the boy sure enough !" she gasped.

"I moughtn't be able ter holp it; I dun'no' how
the law stands. He air kin ter Lee-yander, an'
mebbe hev got the bes' right ter him."

She shivered slightly ; the dew was falling, and
all the budding herbage was glossed with a silver
glister. The shadows were sparse. The white
branches of the aspens cast only the symmetrical
outline of the tree form on the illumined grass, and
seemed scarcely less bare than in winter, but on one
swaying bough the mocking-bird sang all the joyous
prophecies of the spring to the great silver moon
that made his gladsome day so long.

She was quick to notice the sudden cessation of
his song, the alert, downward poise of his beautiful
head, his tense critical attitude. A mimicking whistle rose on the air, now soft, now keen, with swift
changes and intricate successions of tones, ending
in a brilliant borrowed roulade, delivered with a
wonderful velocity and *élan*. The long tail feath-
ers, all standing stiffly upward, once more drooped;
the mocking-bird turned his head from side to side,
then lifting his full throat he poured forth again his
incomparable, superb, infinitely versatile melody,
fixing his glittering eye on the moon, and heeding
the futilely ambitious worldling no more.

The mimicking sound heralded the approach of
Leander. Laurelia's heart, full of bitterness for
his sake, throbbed tenderly for him. Ah, what was

to be his fate! What unkind lot did the future hold for him in the clutches of a man like this! Suddenly she was pitying his mother—her own children, how safe!

She winced to tell him what had happened, but she it was who, bracing her nerves, made the disclosure, for Sudley remained silent, the end of the ox-yoke in his trembling hands, his head bare to the moon and the dew, his face grown lined and old.

Leander stood staring at her out of his moonlit blue eyes, his hat far back on the brown curls she had so vaunted, damp and crisp and clinging, the low limp collar of his unbleached shirt showing his round full throat, one hand resting on the high curb of the well, the other holding a great brown gourd full of the clear water which he had busied himself in securing while she sought to prepare him to hear the worst. His lips, like a bent bow as she thought, were red and still moist as he now and then took the gourd from them, and held it motionless in the interest of her narration, that indeed touched him so nearly. Then, as she made point after point clear to his comprehension, he would once more lift the gourd and drink deeply, for he had had an active day, inducing a keen thirst.

She had been preparing herself for the piteous spectacle of his frantic fright, his futile reliance on them who had always befriended him, his callow forlorn helplessness, his tears, his reproaches; she dreaded them.

He was silent for a reflective moment when she had paused. "But what's he want with me, Cap'n?" he suddenly demanded. "Mought know

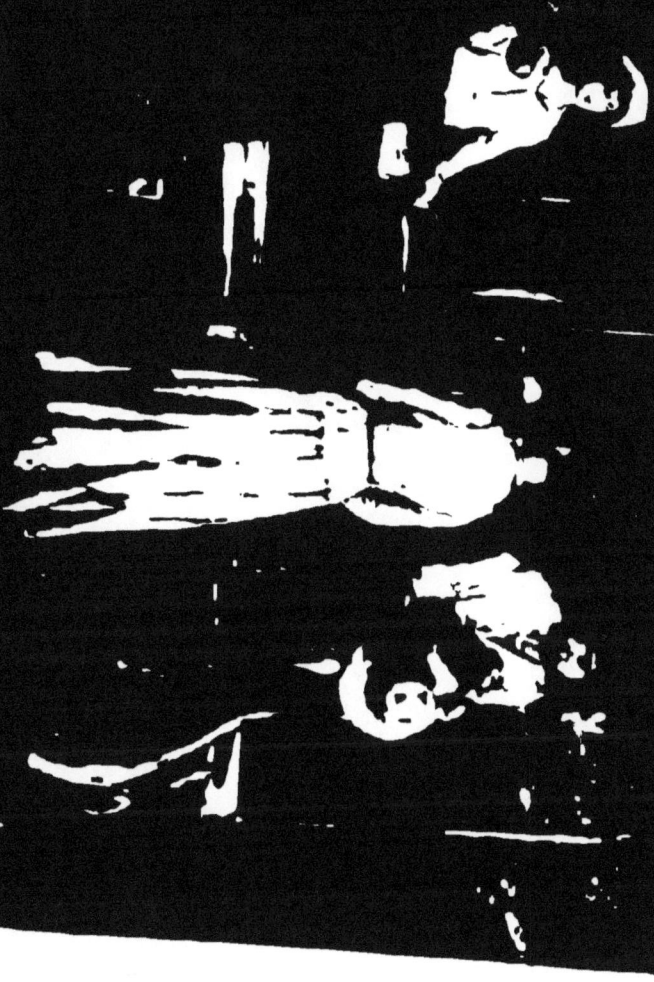

"HE HAD HAD AN ACTIVE DAY, INDUCING A KEEN THIRST"

I warn't industrious in the field, ez he seen me off a-fiddlin' in the woods whilst Neighbor war a-ploughin'."

"Mebbe he 'lows he mought *make* ye industrious an' git cornsider'ble work out'n ye," she faltered, flinching for him.

After another refreshing gulp from the gourd he canvassed this dispassionately. "Say his own chil'n air ' hearty feeders an' hard on shoe-leather ?' Takes a ·good deal o' goadin' ter git ploughin' enough fur the wuth o' feed out'n a toler'ble beastis like old Blaze-face thar, don't it, Neighbor ?—an' how is it a-goin' ter be with a human ez mebbe will hold back an' air sot agin ploughin' ennyhow, an' air sorter idle by profession ? 'Twould gin him a heap o' trouble—more'n the ploughin' an'.sech would be wuth — a heap o' trouble." Once more he bowed his head to the gourd.

"He 'lowed ye shouldn't dwell no mo' in the tents o' sin. He seen the fiddle, Lee ; it's all complicated with the fiddle," she quavered, very near tears of vexation.

He lifted a smiling moonlit face ; his half-suppressed laugh echoed gurglingly in the gourd. "Cap'n," he said, reassuringly, "jes let's hear Uncle Nehemiah talk some mo', an' ef I can't see no mo' likely work fur me 'n ploughin', I'll think myself mighty safe."

They felt like three conspirators as after supper they drew their chairs around the fire with the unsuspicious Uncle Nehemiah. However, Nehemiah Yerby could hardly be esteemed unsuspicious in any point of view, so full of vigilant craft was his

intention in every anticipation, so slyly sanctimo-
nious was his long countenance.

There could hardly have been a greater contrast
than Tyler Sudley's aspect presented. His candid face seemed a mirror for his thought; he had
had scant experience in deception, and he proved
a most unlikely novice in the art. His features
were heavy and set; his manner was brooding and
depressed; he did not alertly follow the conversation; on the contrary, he seemed oblivious of it as
his full dark eyes rested absently on the fire. More
than once he passed his hand across them with a
troubled, harassed manner, and he sighed heavily.
For which his co-conspirators could have fallen
upon him. How could he be so dull, so forgetful
of all save the fear of separation from the boy
whom he had reared, whom he loved as his own
son; how could he fail to know that a jaunty, assured mien might best serve his interests until at
any rate the blow had fallen; why should he wear
the insignia of defeat before the strength of his
claim was tested? Assuredly his manner was calculated to greatly reinforce Nehemiah Yerby's confidence, and to assist in eliminating difficulties in
the urging of his superior rights and the carrying out of his scheme. Mrs. Sudley's heart sank
as she caught a significant gleam from the boy's
eyes; he too appreciated this disastrous policy,
this virtual surrender before a blow was struck.

"An' Ty ain't afeard o' bars," she silently commented, "nor wolves, nor wind, nor lightning, nor
man in enny kind o' a free fight; but bekase he
dun'no' how the *law* stands, an' air afeard the law

mought be able ter take Lee-yander, he jes sets thar ez pitiful ez a lost kid, fairly ready ter blate aloud."

She descried the covert triumph twinkling among the sparse light lashes and "crow-feet" about Nehemiah's eyes as he droned on an ever-lengthening account of his experiences since leaving the county.

" It's a mighty satisfyin' thing ter be well off in yearthly goods an' chattels," said Laurelia, with sudden inspiration. " Ty, thar, is in debt."

For Uncle Nehemiah had been dwelling unctuously upon the extent to which it had pleased the Lord to prosper him. His countenance fell suddenly. His discomfiture in her unexpected disclosure was twofold, in that it furnished a reason for Tyler's evident depression of spirits, demolishing the augury that his manner had afforded as to the success of the guest's mission, and furthermore, to Nehemiah's trafficking soul, it suggested that a money consideration might be exacted to mollify the rigors of parting.

For Nehemiah Yerby had risen to the dignities, solvencies, and responsibilities of opening a store at the cross-roads in Kildeer County. It was a new and darling enterprise with him, and his mind and speech could not long be wiled away from the subject. This abrupt interjection of a new element into his cogitations gave him pause, and he did not observe the sudden rousing of Tyler Sudley from his revery, and the glance of indignant reproach which he cast on his wife. No man, however meek, or however bowed down with sorrow,

will bear unmoved a gratuitous mention of his
debts; it seems to wound him with all the rancor
of insult, and to enrage him with the hopelessness
of adequate retort or reprisal. It is an indignity, .
like taunting a ghost with cock-crow, or exhorting
a clergyman to repentance. He flung himself all
at once into the conversation, to bar and baffle any
renewed allusion to that subject, and it was accident rather than intention which made him grasp
Nehemiah in the vise of a quandary also.

"Ye say ye got a store an' a stock o' truck, Nehemiah. Air ye ekal ter keepin' store an' sech?"
he demanded, speculatively, with an inquiring and
doubtful corrugation of his brows, from which a
restive lock of hair was flung backward like the
toss of a horse's mane.

"I reckon so," Nehemiah sparely responded,
blinking at him across the fireplace.

"An' ye say ye hev applied fur the place o' postmaster?" Tyler prosed on. "All that takes a
power o' knowledge—readin' an' writin' an' cipher-
in' an' sech. How air ye expectin' to hold out,
'kase I know ye never hed no mo' larnin' than me,
an' I war acquainted with ye till ye war thirty years
old an' better?"

The tenor of this discourse did not comport with
his customary suavity and tactful courtesy toward
a guest, but he was much harassed and had lost
his balance. He had a vague idea that Mrs. Sudley hung upon the flank of the conversation with a
complete summary of amounts, dates, and names
of creditors, and he sought to balk this in its in-
ception. Moreover, his forbearance with Nehe-

miah, with his presence, his personality, his mission, had begun to wane. Bitter reflections might
suffice to fill the time were he suffered to be silent;
but since a part in the conversation had been made
necessary, he had for it no honeyed words.

"I'd make about ez fit a postmaster, I know, ez
that thar old *owel* a-hootin' out yander. I could
look smart an' sober like him, but that's 'bout all
the fur my school-larnin' kerried me, an' yourn
didn't reach ter the nex' mile-post — an' that I
know."

Nehemiah's thin lips seemed dry. More than
once his tongue appeared along their verges as he
nervously moistened them. His small eyes had
brightened with an excited look, but he spoke very
slowly, and to Laurelia it seemed guardedly.

"I tuk ter my book arterward, Brother Sudley.
I applied myself ter larnin' vigorous. Bein' ez I
seen the Lord's hand war liberal with the gifts o'
this worl', I wanted ter stir myself ter desarve the
good things."

Sudley brought down the fore-legs of his chair
to the floor with a thump. Despite his anxiety a
slow light of ridicule began to kindle on his face;
his curling lip showed his strong white teeth.

"Waal, by gum! ye mus' hev been a sight ter be
seen! Ye, forty or fifty years old, a-settin' on the
same seat with the chil'n at the deestric' school, an'
a-competin' with the leetle tadpoles fur 'Baker an'
Shady' an sech!"

He was about to break forth with a guffaw of
great relish when Nehemiah spoke hastily, fore-
stalling the laughter.

"Naw; Abner Sage war thar fur a good while las' winter a - visitin' his sister, an' he kem an' gin me lessons an' set me copies thar at my house, an' I larnt a heap."

Leander lifted his head suddenly. The amount of progress possible to this desultory and limited application he understood only too well. He had not learned so much himself to be unaware how much in time and labor learning costs. The others perceived no incongruity. Sudley's face was florid with pride and pleasure, and his wife's reflected the glow.

"Ab Sage at the cross-roads! Then he mus' hev tole ye 'bout Lee-yander hyar, an' his larnin'. Ab tole, I know."

Nehemiah drew his breath in quickly. His twinkling eyes sent out the keenest glance of suspicion, but the gay, affectionate, vaunting laugh, as Tyler Sudley turned around and clapped the boy a ringing blow on his slender shoulder, expressed only the plenitude of his simple vainglory.

"Lee-yander hyar *knows it all!*" he boasted. "Old Ab himself don't know no mo'! I'll be bound old Ab went a-braggin'—hey, Lee-yander?"

But the boy shrank away a trifle, and his smile was mechanical as he silently eyed his relative.

"Ab 'lowed he war tur'ble disobejient," said Nehemiah, after a pause, and cautiously allowing himself to follow in the talk, "an' gi'n over ter playin' the fiddle." He hesitated for a moment, longing to stigmatize its ungodliness; but the recollection of Tyler Sudley's uncertain temper decided him, and he left it unmolested. "But Ab 'lowed ye war middlin' quick at figgers, Lee-yander—middlin' quick at figgers!"

Leander, still silent and listening, flushed slightly. This measured praise was an offence to him; but he looked up brightly and obediently when his uncle wagged an uncouthly sportive head (Nehemiah's anatomy lent itself to the gay and graceful with much reluctance), thrust his hands into his pockets, and, tilting himself back in his chair, continued:

"I'll try ye, sonny—I'll try ye. How much air nine times seven?—nine times seven?"

"Forty-two!" replied the boy, with a bright, docile countenance fixed upon his relative.

There was a pause. "Right!" exclaimed Nehemiah, to the relief of Sudley and his wife, who had trembled during the pause, for it seemed so threatening. They smiled at each other, unconscious that the examination meant aught more serious than a display of their prodigy's learning.

"An', now, how much air twelve times eight?" demanded Nehemiah.

"Sixty-six!" came the answer, quick as lightning.

"Right, sir, every time!" cried Nehemiah with a glow of genuine exultation, as he brought down the fore-legs of the chair to the floor, and the two Sudleys laughed aloud with pleasure.

Leander saw them all distorted and grimacing while the room swam round. The scheme was clear enough to him now. The illiterate Nehemiah, whose worldly prosperity had outstripped his mental qualifications, had bethought himself of fill-

ing the breach with his nephew, given away as
surplusage in his burdensome infancy, but trans-
formed into a unique utility under the tutelage of
Abner Sage. It was his boasting of his froward
pupil, doubtless, that had suggested the idea, and
Leander understood now that he was to do the
work of the store and the post-office under the
nominal incumbency of this unlettered lout. Had
the whole transaction been open and acknowl-
edged, Leander would have had scant appetite for
the work under this master; but he revolted at the
flimsy, contemptible sham; he bitterly resented the
innuendoes against the piety of the Sudleys, not
that he cared for piety, save in the abstract; he
was daunted by the brutal ignorance, the doltish
inefficiency of the imposture that had so readily
accepted his patently false answers to the simple
questions. He had a sort of crude reverence for
education, and it had seemed to him a very serious
matter to take such liberties with the multiplication table. He valued, too, with a boy's stalwart
vanity, his reputation for great learning, and he
would not have lightly jeopardized it did he not
esteem the crisis momentous. He knew not what
he feared. The fraud of the intention, the ground-
less claim to knowledge, made Nehemiah's scheme
seem multifariously guilty in some sort; while
Tyler Sudley and his wife, albeit no wiser mathe-
matically, had all the sanctions of probity in their
calm, unpretending ignorance.

"Ef Cap'n or Neighbor wanted ter run a post-
office on my larnin', or ter keep store, they'd be
welcome; but I won't play stalkin'-horse fur that

thar man's still-hunt, sure ez shootin'," he said to himself.

The attention which he bent upon the conversation thenceforth was an observation of its effect rather than its matter. He saw that he was alone in his discovery. Neither Sudley nor his wife had perceived any connection between the store, the prospective post-office, and the desire of the illiterate would-be postmaster to have his erudite nephew restored to his care.

It may be that the methods of his "Neighbor" and the "Captain" in the rearing of Leander, the one with unbridled leniency, the other with spurious severity and affected indifference, had combined to foster self-reliance and decision of character, or it may be that these qualities were inherent traits.. At all events, he encountered the emergency without an instant's hesitation. He felt no need of counsel. He had no doubts. He carried to his pallet in the roof-room no vacillations and no problems. His resolve was taken. For a time, as he listened to the movements below-stairs, the sound of voices still rose, drowsy as the hour waxed late; the light that flickered through the cracks in the puncheon flooring gradually dulled, and presently a harsh grating noise acquainted him with the fact that Sudley was shovelling the ashes over the embers; then the tent-like attic was illumined only by the moonlight admitted through the little square window at the gable end—so silent, so still, it seemed that it too slept like the silent house. The winds slumbered amidst the mute woods; a bank of cloud that he could see from

his lowly couch lay in the south becalmed. The
bird's song had ceased. It seemed to him as he
lifted himself on his elbow that he had never known
the world so hushed. The rustle of the quilt of
gay glazed calico was of note in the quietude; the
impact of his bare foot on the floor was hardly a
sound, rather an annotation of his weight and his
movement; yet in default of all else the sense of
hearing marked it. His scheme seemed impracticable as for an instant he wavered at the head of the
ladder that served as a stairway; the next moment
his foot was upon the rungs, his light, lithe figure
slipping down it like a shadow. The room below,
all eclipsed in a brown and dusky-red medium, the
compromise between light and darkness that the
presence of the embers fostered, was vaguely revealed to him. He was hardly sure whether he saw
the furniture all in place, or whether he knew its
arrangement so well that he seemed to see. Suddenly, as he laid his hand on the violin on the wall,
it became visible, its dark red wood richly glowing
against the brown logs and the tawny clay daubing.
A tiny white flame had shot up in the midst of the
gray ashes, as he stood with the cherished object
in his cautious hand, his excited eyes, dilated and
expectant, searching the room apprehensively, while
a vague thrill of a murmur issued from the instrument, as if the spirit of music within it had been
wakened by his touch—too vague, too faintly elusive for the dormant and somewhat dull perceptions
of Nehemiah Yerby, calmly slumbering in state in
the best room.

The faint jet of flame was withdrawn in the ashes

as suddenly as it had shot forth, and in the ensuing
darkness, deeper for the contrast with that momentary illumination, it was not even a shadow that
deftly mounted the ladder again and emerged into
the sheeny twilight of the moonlit roof-room. Leander was somehow withheld for a moment motionless
at the window; it may have been by compunction;
it may have been by regret, if it be possible to the
very young to definitely feel either.   There was an
intimation of pensive farewell in his large illumined eyes as they rested on the circle of familiar things about him—the budding trees, the well,
with its great angular sweep against the sky, the
still sward, the rail-fences glistening with the dew,
the river with the moonlight in a silver blazonry
on its lustrous dark surface, the encompassing
shadows of the gloomy mountains.   There was no
sound, not even among the rippling shallows; he
could hear naught but the pain of parting throbbing in his heart, and from the violin a faint continuous susurrus, as if it murmured half-asleep
memories of the melodies that had thrilled its
waking moments.   It necessitated careful hand-
ling as he deftly let himself out of the window,
the bow held in his mouth, the instrument in one
arm, while the other hand clutched the boughs of
a great holly-tree close beside the house.   It was
only the moonlight on those smooth, lustrous leaves,
but it seemed as if smiling white faces looked
suddenly down from among the shadows: at this
lonely hour, with none awake to see, what strange
things may there not be astir in the world, what un-
measured, unknown forces, sometimes felt through

15

the dulling sleep of mortals, and then called dreams!
As he stood breathless upon the ground the wind
awoke. He heard it race around the corner of the
house, bending the lilac bushes, and then it softly
buffeted him full in the face and twirled his hat on
the ground. As he stooped to pick it up he heard
whispers and laughter in the lustrous boughs of
the holly, and the gleaming faces shifted with the
shadows. He looked fearfully over his shoulder; the rising wind might waken some one of the
household. His "Neighbor" was, he knew, solici-
tous about the weather, and suspicious of its intentions lest it not hold fine till all the oats be
sown. A pang wrung his heart; he remembered
the long line of seasons when, planting corn in the
pleasant spring days, his "Neighbor" had opened
the furrow with the plough, and the "Captain"
had followed, dropping the grains, and he had
brought up the rear with his hoe, covering them
over, while the clouds floated high in the air, and
the mild sun shone, and the wind kept the shadows
a-flicker, and the blackbird and the crow, com-
placently and craftily watching them from afar,
seemed the only possible threatening of evil in all
the world. He hastened to stiffen his resolve.
He had need of it. Tyler Sudley had said that
he did not know how the law stood, and for himself, he was not willing to risk his liberty on it.
He gazed apprehensively upon the little batten
shutter of the window of the room where Nehemiah Yerby slept, expecting to see it slowly swing
open and disclose him there. It did not stir, and
gathering resolution from the terrors that had be-

set him when he fancied his opportunity threatened, he ran like a frightened deer fleetly down the road, and plunged into the dense forest. The wind kept him company, rollicking, quickening, coming and going in fitful gusts. He heard it die away, but now and again it was rustling among a double file of beech-trees all up the mountainside. He saw the commotion in their midst, the effect of swift movement as the scant foliage fluttered, then the white branches of the trees all a-swaying like glistening arms flung upward, as if some bevy of dryads sped up the hill in elusive rout through the fastnesses.

The next day ushered in a tumult and excitement unparalleled in the history of the little log-cabin. When Leander's absence was discovered, and inquiry of the few neighbors and search of the vicinity proved fruitless, the fact of his flight and its motive were persistently forced upon Nehemiah Yerby's reluctant perceptions, with the destruction of his cherished scheme as a necessary sequence. With some wild craving for vengeance he sought to implicate Sudley as accessory to the mysterious disappearance. He found some small measure of solace in stumping up and down the floor before the hearth, furiously railing at the absent host, for Sudley had not yet relinquished the bootless quest, and indignantly upbraiding the forlorn, white-faced, grief-stricken Laurelia, who sat silent and stony, her faded eyes on the fire, heedless of his words. She held in her lap sundry closely-rolled knitted balls—the boy's socks that she had so carefully made and darned. A pile of his clothing lay at her feet. He had carried nothing but his fiddle and the clothes he stood in, and if she had had more tears she could have wept for his improvidence, for the prospective tatters and rents that must needs befall him in that unknown patchless life to which he had betaken himself.

Nehemiah Yerby argued that it was Sudley who had prompted the whole thing; he had put the boy up to it, for Leander was not so lacking in feeling as to flee from his own blood-relation. But he would set the law to spy them out. He would be back again, and soon.

He may have thought better of this presently, for he was in great haste to be gone when Tyler Sudley returned, and to his amazement in a counterpart frame of mind, charging Nehemiah with the responsibility of the disaster. It was strange to Laurelia that she, who habitually strove to fix her mind on religious things, should so relish the aspect of Ty Sudley in his secular rage on this occasion.

"Ye let we-uns hev him whilst so leetle an' helpless, but now that he air so fine growed an' robustious ye want ter git some work out'n him, an' he hev runned away an' tuk ter the woods tarrified by the very sight of ye," he averred. "He'll never kem back; no, he'll never kem back; fur he'll 'low ez ye would kem an' take him home with you; an' now the Lord only knows whar he is, an' what will become of him."

His anger and his tumultuous grief, his wild, irrepressible anxiety for Leander's safety, convinced the crafty Nehemiah that he was no party to the boy's scheme. Sudley's sorrow was not of the kind that renders the temper pliable, and when Nehemiah sought to point a moral in the absence of the violin, and for the first time in Sudley's presence protested that he desired to save Leander from that device of the devil, the master of the house shook his inhospitable fist very close indeed to his guest's nose, and Yerby was glad enough to follow that feature unimpaired out to his horse at the bars, saying little more.

He aired his views, however, at each house where he made it convenient to stop on his way home, and took what comfort there might be in the rôle of martyr. Leander was unpopular in several localities, and was esteemed a poor specimen of the skill of the Sudleys in rearing children. He had been pampered and spoiled, according to general report, and more than one of his successive interlocutors were polite enough to opine that the change to Nehemiah's charge would have been a beneficent opportunity for much-needed discipline. Nehemiah was not devoid of some skill in interrogatory. He contrived to elicit speculations without giving an intimation of unduly valuing the answer.

" He's 'mongst the moonshiners, I reckon," was the universal surmise. " He'll be hid mighty safe 'mongst them."

For where the still might be, or who was engaged in the illicit business, was even a greater mystery than Leander's refuge. Nothing more definite could be elicited than a vague rumor that some such work was in progress somewhere along the many windings of Hide-and-Seek Creek.

Nehemiah Yerby had never been attached to temperance principles, and, commercially speaking, he had thought it possible that whiskey on which no tax had been paid might be more profitably dispensed at his store than that sold under the sanctions of the government. These considerations, however, were as naught in view of the paralysis which his interests and schemes had suffered in Leander's flight. He dwelt with dismay upon the possibility that he might secure the postmastership without the capable assistant whose services were essential. In this perverse sequence of events disaster to his application was more to be desired than success. He foresaw himself browbeaten, humiliated, detected, a butt for the ridicule of the community, his pretensions in the dust, his pitiful imposture unmasked. And beyond these æsthetic misfortunes, the substantial emoluments of "keepin' store," with a gallant sufficiency of arithmetic to regulate prices and profits, were vanishing like the elusive matutinal haze before the noontide sun. Nehemiah Yerby groaned aloud, for the financial stress upon his spirit was very like physical pain. And in this inauspicious moment he bethought himself of the penalties of violating the Internal Revenue Laws of the United States.

Now it has been held by those initiated into such mysteries that there is scant affinity between whiskey and water. Nevertheless, in this connection, Nehemiah Yerby developed an absorbing interest in the watercourses of the coves and adjacent

mountains, especially their more remote and sequestered tributaries. He shortly made occasion
to meet the county surveyor and ply him with questions touching the topography of the vicinity, cloak-
ing the real motive under the pretence of an interest in water-power sufficient and permanent enough
for the sawing of lumber, and professing to contemplate the erection of a saw-mill at the most
eligible point. The surveyor had his especial vanity, and it was expressed in his frequent boast that
he carried a complete map of the county graven
upon his brain; he was wont to esteem it a gracious opportunity when a casual question in a
group of loungers enabled him to display his familiarity with every portion of his rugged and moun-
tainous region, which was indeed astonishing, even
taking into consideration his incumbency for a
number of terms, aided by a strong head for locality. Nehemiah Yerby's scheme was incalculably
favored by this circumstance, but he found it unexpectedly difficult to support the figment which
he had propounded as to his intentions. Fiction
is one of the fine arts, and a mere amateur like
Nehemiah is apt to fail in point of consistency.
He was inattentive while the surveyor dilated on
the probable value, the accessibility, and the relative height of the "fall" of the various sites, and
their available water-power, and he put irrelevant
queries concerning ineligible streams in other localities. No man comfortably mounted upon his
hobby relishes an interruption. The surveyor would
stop with a sort of bovine surprise, and break out
in irritable parenthesis.

"That branch on the t'other side o' Panther Ridge? Why, man alive, that thread o' water wouldn't turn a spider web."

Nehemiah, quaking under the glance of his keen questioning eye, would once more lapse into silence, while the surveyor, loving to do what he could do well, was lured on in his favorite subject by the renewed appearance of receptivity in his listener.

"Waal, ez I war a-sayin', I know every furlong o' the creeks once down in the Cove, an' all their meanderings, an' the best part o' them in the hills amongst the laurel and the wildernesses. But now the ways of sech a stream ez Hide-an'-Seek Creek are past finding out. It's a 'sinking creek,' you know; goes along with a good volume and a swift current for a while to the west, then disappears into the earth, an' ain't seen fur five mile, then comes out agin running due north, makes a tremenjious jump—the Hoho-hebee Falls—then pops into the ground agin, an' ain't seen no more forever," he concluded, dramatically.

"How d'ye know it's the same creek?" demanded Nehemiah, sceptically, and with a wrinkling brow.

"By settin' somethin' afloat on it before it sinks into the ground—a piece of marked bark or a shingle or the like—an' finding it agin after the stream comes out of the caves," promptly replied the man of the compass, with a triumphant snap of the eye, as if he entertained a certain pride in the vagaries of his untamed mountain friend. "Nobody knows how often it disappears, nor where it

rises, nor where it goes· at last.   It's got dozens of
fust-rate millin' sites, but then it's too fur off fur
you ter think about."

"Oh no 'tain't!" exclaimed Nehemiah, suddenly.

The surveyor stared.   "Why, you ain't thinkin'
'bout movin' up inter the wilderness ter live, an' ye
jes applied fur the post-office down at the cross-
roads?   Ye can't run the post-office thar an' a saw-
mill thirty mile away at the same time."

Nehemiah was visibly disconcerted. His wrinkled
face showed the flush of discomfiture, but his craft
rallied to the emergency.

"Moughtn't git the post-office, arter all's come
an' gone.   Nothin' is sartin in this vale o' tears."

"An' ye air goin' ter take ter the woods ef ye
don't?" demanded the surveyor, incredulously.
"Thought ye war goin' ter keep store?"

"Waal, I dun'no'; jes talkin' round," said Nehemiah, posed beyond recuperation.   "I mus' be
a-joggin', ennyhow.   Time's a-wastin'."

As he made off hastily in the direction of his
house, for this conversation had taken place at the
blacksmith's shop at the cross-roads, the surveyor
gazed after him much mystified.

"What is that old fox slyin' round after? He
ain't studyin' 'bout no saw-mill, inquirin' round
about all the out-o'-the-way water-power in the kentry fifty mile from where he b'longs. He's a heap
likelier to be goin' ter start a wild-cat still in them
wild places—git his whiskey cheap ter sell in his
store."

He shook his head sagely once for all, for the
surveyor's mind was of the type prompt in reach-

ing conclusions, and he was difficult to divert from his convictions.

A feature of the development of craft to a certain degree is the persuasion that this endowment is not shared. A fine world it would be if the Nehemiah Yerbys were as clever as they think themselves, and their neighbors as dull. He readily convinced himself that he had given no intimation that his objects and motives were other than he professed, and with unimpaired energy he went to work upon the lines which he had marked out for himself. A fine chase Hide-and-Seek Creek led him, to be sure, and it tried his enthusiasms to the uttermost. What affinity this brawling vagrant had for the briers and the rocks and the tangled fastnesses! Seldom, indeed, could he press in to its banks and look down upon its dimpled, laughing, heedless face without the sacrifice of fragments of flesh and garments left impaled upon the sharp spikes of the budding shrubs. Often it so intrenched itself amidst the dense woods, and the rocks and chasms of its craggy banks, that approach was impossible, and he followed it for miles only by the sound of its wild, sweet, woodland voice. And this, too, was of a wayward fancy; now, in turbulent glee among the rocks, riotously chanting aloud, challenging the echoes, and waking far and near the forest quiet; and again it was merely a low, restful murmur, intimating deep, serene pools and a dallying of the currents, lapsed in the fulness of content. Then Nehemiah Yerby would be beset with fears that he would lose this whisper, and his progress was slight; he would pause to listen, hearing nothing; would turn to right, to left; would take his way back through the labyrinth of the laurel to catch a thread of sound, a mere crystalline tremor, and once more follow this transient lure. As the stream came down a gorge at a swifter pace and in a succession of leaps—a glassy cataract visible here and there, airily sporting with rainbows, affiliating with ferns and moss and marshy growths, the bounding spray glittering in the sunshine—it flung forth continuously tinkling harmonies in clear crystal tones, so penetrating, so definitely melodic, that more than once, as he paced along on his jaded horse, he heard in their midst, without disassociating the sounds, the *ping, pang, ping, pang,* of the violin he so condemned. He drew up at last, and strained his ear to listen. It did not become more distinct, always intermingled with the recurrent rhythm of the falling water, but always vibrating in subdued throbbings, now more acute, now less, as the undiscriminated melody ascended or descended the scale. It came from the earth, of this he was sure, and thus he was reminded anew of the caves which Hide-and-Seek Creek threaded in its long course. There was some opening near by, doubtless, that led to subterranean passages, dry enough here, since it was the stream's whim to flow in the open sunshine instead of underground. He would have given much to search for it had he dared. His leathery, lean, loose cheek had a glow of excitement upon it; his small eyes glistened; for the first time in his life, possibly, he looked young. But he did not doubt that this was the stronghold of the illicit distillers, of whom one heard so much in the Cove and saw so little. A lapse of caution, an inconsiderate movement, and he, might be captured and dealt with as a spy and informer.

Nevertheless his discovery was of scant value unless he utilized it further. He had always believed that his nephew had fled to the secret haunts of the moonshiners. Now he only knew it the more surely; and what did this avail him, and how aid in the capture of the recusant clerk and assistant postmaster? He hesitated a moment; then fixing the spot in his mind by the falling of a broad crystal sheet of water from a ledge some forty feet high, by a rotting log at its base that seemed to rise continually, although the moving cataract appeared motionless, by certain trees and their relative position, and the blue peaks on a distant skyey background of a faint cameo yellow, he slowly turned his horse's rein and took his way out of danger. It was chiefly some demonstration on the animal's part that he had feared. A snort, a hoof-beat, a whinny would betray him, and very liable was the animal to any of these expressions. One realizes how unnecessary is speech for the exposition of opinion when brought into contradictory relations with the horse which one rides or drives. All day had this animal snorted his doubts of his master's sanity; all day had he protested against these aimless, fruitless rambles; all day had he held back with a high head and a hard mouth, while whip and spur pressed him through laurel almost impenetrable, and through crevices of crags almost impassable. For were there not all the fair roads

of the county to pace and gallop upon if one must
needs be out and jogging! Unseen objects, vaguely discerned to be moving in the undergrowth affrighted the old plough-horse of the levels—infinitely reassured and whinnying with joyful relief when
the head of horned cattle showed presently as the
cause of the commotion. He would have given
much a hundred times that day, and he almost said
so a hundred times, too, to be at home, with the
old bull-tongue plough behind him, running the
straight rational furrow in the good bare open field,
so mellow for corn, lying in the sunshine, inviting
planting.

"Ef I git ye home wunst more, I'll be bound I'll
leave ye thar," Nehemiah said, ungratefully, as they
wended their way along; for without the horse he
could not have traversed the long distances of his
search, however unwillingly the aid was given.

He annotated his displeasure by a kick in the
ribs; and when the old equine farmer perceived that
they were absolutely bound binward, and that their
aberrations were over for the present, he struck a
sharp gait that would have done honor to his youthful days, for he had worn out several pairs of legs
in Nehemiah's fields, and was often spoken of as
being upon the last of those useful extremities. He
stolidly shook his head, which he thought so much
better than his master's, and bedtime found them
twenty miles away and at home.

Nehemiah felt scant fatigue. He was elated with
his project. He scented success in the air. It
smelled like the season. It too was suffused with
the urgent pungency of the rising sap, with the fra-

grance of the wild-cherry, with the vinous promise of the orchard, with the richness of the mould, with the vagrant perfume of the early flowers.

He lighted a tallow dip, and he sat him down with writing materials at the bare table to indite a letter while all his household slept. The windows stood open to the dark night, and Spring hovered about outside, and lounged with her elbows on the sill, and looked in. He constantly saw something pale and elusive against the blackness, for there was no moon, but he thought it only the timid irradiation with which his tallow dip suffused the blossoming wands of an azalea, growing lithe and tall hard by. With this witness only he wrote the letter—an anonymous letter, and therefore he was indifferent to the inadequacies of his penmanship and his spelling. He labored heavily in its composition, now and then perpetrating portentous blots. He grew warm, although the fire that had served to cook supper had long languished under the bank of ashes. The tallow dip seemed full of caloric, and melted rapidly in pendulous drippings. He now and again mopped his red face, usually so bloodless, with his big bandanna handkerchief, while all the zephyrs were fanning the flying tresses of Spring at the window, and the soft, sweet, delicately attuned vernal chorus of the marshes were tentatively running over *sotto voce* their allotted melodies for the season. Oh, it was a fine night outside, and why should a moth, soft-winged and cream-tinted and silken-textured, come whisking in from the dark, as silently as a spirit, to supervise Nehemiah Yerby's letter, and travel up and down the

page all befouled with the ink? And as he sought
to save the sense of those significant sentences from
its trailing silken draperies, why should it rise suddenly, circling again and again about the candle,
pass through the flame, and fall in quivering agonies
once more upon the page? He looked at it, dead
now, with satisfaction. It had come so very near
ruining his letter—an important letter, describing
the lair of the illicit distillers to a deputy marshal
of the revenue force, who was known to be in a
neighboring town. He had good reason to with-
hold his signature, for the name of the informer in
the ruthless vengeance of the region would be as
much as his life was worth. The moth had not
spoiled the letter—the laborious letter; he was so
glad of that! He saw no analogies, he received
not even a subtle warning, as he sealed and addressed the envelope and affixed the postage-stamp.
Then he snuffed out the candle with great satisfaction.

The next morning the missive was posted, and
all Nehemiah Yerby's plans took a new lease of
life. The information he had given would result
in an immediate raid upon the place. Leander
would be captured among the moonshiners, but his
youth and his uncle's representations—for he would
give the officers an inkling of the true state of the
case—would doubtless insure the boy's release,
and his restoration to those attractive commercial
prospects which had been devised for him.

The ordering of events is an intricate process, and to its successful exploitation a certain degree of sagacious prescience is a prerequisite, as well as a thorough mastery of the lessons of experience. For a day or so all went well in the inner consciousness of Nehemiah Yerby. The letter had satisfied his restless craving for some action toward the consummation of his ambition, and he had not the foresight to realize how soon the necessity of following it up would supervene. He first grew uneasy lest his letter had not reached its destination; then, when the illimitable field of speculation was thus opened out, he developed an ingenuity of imagination in projecting possible disaster. Day after day passed, and he heard naught of his cherished scheme. The revenuers—craven wretches he deemed them, and he ground his teeth with rage because of their seeming cowardice in their duty, since their duty could serve his interests—might not have felt exactly disposed to risk their lives in these sweet spring days, when perhaps even a man whose life belongs to the government might be presumed to take some pleasure in it, by attempting to raid the den of a gang of moonshiners on the scanty faith of an informer's word, tenuous guaranty at best, and now couched in an anony-

mous letter, itself synonym for a lie. Oh, what fine
eulogies rose in his mind upon the manly virtue of
courage! How enthusing it is at all times to contemplate the courage of others!—and how safe!

Then a revulsion of belief ensued, and he began
to fear that they might already have descended
upon their quarry, and with all their captives have
returned to the county town by the road by which
they came—nearer than the route through the cross-
roads, though far more rugged. Why had not this
possibility before occurred to him! He had so
often prefigured their triumphant advent into the
hamlet with all their guarded and shackled prisoners, the callow Leander in the midst, and his own
gracefully enacted rôle of virtuous, grief-stricken,
pleading relative, that it seemed a recollection—
something that had really happened—rather than
the figment of anticipation. But no word, no breath
of intimation, had ruffled the serenity of the cross-
roads. The calm, still, yellow sunshine day by day
suffused the land like the benignities of a dream
—almost too good to be true. Every man with the
heart of a farmer within him was at the plough-
handles, and making the most of the fair weather.
The cloudless sky and the auspicious forecast of
fine days still to come did more to prove to the
farmer the existence of an all-wise, overruling Providence than all the polemics of the world might ac-
complish. The furrows multiplied everywhere save
in Nehemiah's own fields, where he often stood so
long in the turn-row that the old horse would desist
from twisting his head backward in surprise, and
start at last of his own motion, dragging the plough,

16

the share still unanchored in the ground, half across the field before he could be stopped. The vagaries of these "lands" that the absent-minded Nehemiah laid off attracted some attention.

"What ails yer furrows ter run so crooked, Nehemiah?" observed a passer-by, a neighbor who had been to the blacksmith-shop to get his plough-point sharpened ; he looked over the fence critically. "Yer eyesight mus' be failin' some."

"I dun'no'," rejoined Nehemiah, hastily. Then reverting to his own absorption. "War it you-uns ez I hearn say thar war word kem ter the cross-roads 'bout some revenuers. raidin' 'round somewhar in the woods ?"

The look of surprise cast upon him seemed to his alert anxiety to betoken suspicion. "Laws-a-massy, naw !" exclaimed his interlocutor. "Ye air the fust one that hev named sech ez that in these diggin's, fur I'd hev hearn tell on it, sure, ef thar hed been enny sech word goin' the rounds."

Nehemiah recoiled into silence, and presently his neighbor went whistling on his way. He stood motionless for a time, until the man was well out of sight, then he began to hastily unhitch the plough-gear. His resolution was taken. He could wait no longer. For aught he knew the raiders might have come and gone, and be now a hundred miles away with their prisoners to stand their trial in the Federal court. His schemes might have all gone amiss, leaving him in naught the gainer. He could rest in uncertainty no more. He feared to venture further questions when no rumor stirred the air. They rendered him doubly liable to suspicion—to the law-abiding as a possible moonshiner, to any sympathizer with the distillers as a probable informer.   He determined to visit the spot, and there judge how the enterprise had fared.

When next he heard that fine sylvan symphony of the sound of the falling water—the tinkling bell-like tremors of its lighter tones mingling with the sonorous, continuous, deeper theme rising from its weight and volume and movement; with the surging of the wind in the pines; with the occasional cry of a wild bird deep in the new verdure of the forests striking through the whole with a brilliant, incidental, detached effect—no faint vibration was in its midst of the violin's string, listen as he might. More than once he sought to assure himself that he heard it, but his fancy failed to respond to his bidding, although again and again he took up his position where it had before struck his ear.   The wild minstrelsy of the woods felt no lack, and stream and wind and harping pine and vagrant bird lifted their voices in their wonted strains.   He could hardly accept the fact; he would verify anew the landmarks he had made and again return to the spot, his hat in his hand, his head bent low, his face lined with anxiety and suspense.   No sound, no word, no intimation of human presence. The moonshiners were doubtless all gone long ago, betrayed into captivity, and Leander with them. He had so hardened his heart toward his recalcitrant young kinsman and his Sudley friends, he felt so entirely that in being among the moonshiners Leander had met only his deserts in coming to the bar of Federal justice, that he would have experienced scant sorrow if the nephew had not carried off with his own personality his uncle's book-keeper and postmaster's clerk. And so—alas, for Leander! As he meditated on the untoward manner in which he had overshot his target, this marksman of fate forgot the caution which had distinguished his approach, for hitherto it had been as heedful as if he fully believed the lion still in his den. He slowly patrolled the bank below the broad, thin, crystal sheet, seeing naught but its rainbow hovering elusively in the sun, and its green and white skein-like draperies pendulous before the great dark arch over which the cataract fell. The log caught among the rocks in the spray at the base was still there, seeming always to rise while the restless water seemed motionless.

No trace that human beings had ever invaded these solitudes could he discover. No vague, faint suggestion of the well-hidden lair of the moonshiners did the wild covert show forth. "The revenuers war smarter'n me; I'll say that fur 'em," he muttered at last as he came to a stand-still, his chin in his hand, his perplexed eyes on the ground. And suddenly—a footprint on a marshy spot; only the heel of a boot, for the craggy ledges hid all the ground but this, a mere sediment of sand in a tiny hollow in the rock from which the water had evaporated. It was a key to the mystery. Instantly the rugged edges of the cliff took on the similitude of a path. Once furnished with this idea, he could perceive adequate footing all adown the precipitous way. He was not young; his habits had been inactive, and were older even than his age. He could not account for it afterward, but he followed for a few paces this suggestion of a path down the precipitous sides of the stream. He had a sort of triumph in finding it so practicable, and he essayed it still farther, although the sound of the water had grown tumultuous at closer approach, and seemed to foster a sort of responsive turmoil of the senses; he felt his head whirl as he looked at the bounding, frothing spray, then at the long swirls of the current at the base of the fall as they swept on their way down the gorge. As he sought to lift his fascinated eyes, the smooth glitter of the crystal sheet of falling water so close before him dazzled his sight. He wondered afterward how his confused senses and trembling limbs sustained him along the narrow, rugged path, here and there covered with oozing green moss, and slippery with the continual moisture. It evidently was wending to a ledge. All at once the contour of the place was plain to him; the ledge led behind the cataract that fell from the beetling heights above. And within were doubtless further recesses, where perchance the moonshiners had worked their still. As he reached the ledge he could see behind the falling water and into the great concave space which it screened beneath the beetling cliff. It was as he had expected—an arched portal of jagged brown rocks, all dripping with moisture and oozing moss, behind the semi-translucent green-and-white drapery of the cascade.

But he had not expected to see, standing quietly in the great vaulted entrance, a man with his left hand on a pistol in his belt, the mate of which his

more formidable right hand held up with a steady
finger on the trigger.

This much Nehemiah beheld, and naught else,
for the glittering profile of the falls, visible now
only aslant, the dark, cool recess beyond, that
menacing motionless figure at the vanishing-point
of the perspective, all blended together in an indistinguishable whirl as his senses reeled. He
barely retained consciousness enough to throw up
both his hands in token of complete submission.
And then for a moment he knew no more. He
was still leaning motionless against the wall of
rock when he became aware that the man was
sternly beckoning to him to continue his approach.
His dumb lips moved mechanically in response,
but any sound must needs have been futile indeed
in the pervasive roar of the waters. He felt that
he had hardly strength for another step along the
precipitous way, but there is much tonic influence
in a beckoning revolver, and few men are so weak
as to be unable to obey its behests. Poor Nehemiah tottered along as behooved him, leaving all
the world, liberty, volition, behind him as the descending sheet of water fell between him and the
rest of life and shut him off.

"That's it, my leetle man! I thought you could
make it!" were the first words he could distinguish
as he joined the mountaineer beneath the crag.

Nehemiah Yerby had never before seen this man.
That in itself was alarming, since in the scanty popu-
lation of the region few of its denizens are unknown
to each other, at least by sight. The tone of satire,
the gleam of enjoyment in his keen blue eye, were

not reassuring to the object of his ridicule. He was
tall and somewhat portly, and he had a bluff and off-
hand manner, which, however, served not so much
to intimate his good-will toward you as his abound-
ing good-humor with himself. He was a man of most
arbitrary temper, one could readily judge, not only
from his own aspect and manner, but from the docile, reliant, approving cast of countenance of his
reserve force — a half - dozen men, who were somewhat in the background, lounging on the rocks about
a huge copper still. They wore an attentive aspect,
but offered to take no active part in the scene
enacted before them. One of them—even at this
crucial moment Yerby noticed it with a pang of
regretful despair — held noiseless on his knee a
violin, and more than once addressed himself seriously to rubbing rosin over the bow. There was
scant music in his face—a square physiognomy,
with thick features, and a shock of hay-colored hair
striped somewhat with an effect of darker shades
like a weathering stack. He handled the bow with
a blunt, clumsy hand that augured little of delicate
skill, and he seemed from his diligence to think that
rosin is what makes a fiddle play. He was evidently
one of those unhappy creatures furnished with some
vague inner attraction to the charms of music, with
no gift, no sentiment, no discrimination. Something
faintly sonorous there was in his soul, and it vibrated
to the twanging of the strings. He was far less alert
to the conversation than the others, whose listening
attitudes attested their appreciation of the importance of the moment.

"Waal," observed the moonshiner, impatiently,

eying the tremulous and tongue-tied Yerby, " hev ye
fund what ye war a-huntin' fur ?"

So tenacious of impressions was Nehemiah that
it was the violin in those alien hands which still
focussed his attention as he stared gaspingly about.
Leander was not here; probably had never been
here; and the twanging of those strings had lured
him to his fate. Well might he contemn the festive
malevolence of the violin's influence! His letter
had failed; no raider had intimidated these bluff,
unafraid, burly law-breakers, and he had put his
life in jeopardy in his persistent prosecution of his
scheme. He gasped again at the thought.

" *Waal*," said the moonshiner, evidently a man of
short patience, and with a definite air of spurring on
the visitor's account of himself, "we 'ain't been
lookin' fur any spy lately, but I'm 'lowin' ez we hev
fund him."

His fear thus put into words so served to realize
to Yerby his immediate danger that it stood him in
the stead of courage, of brains, of invention; his
flaccid muscles were suddenly again under control;
he wreathed his features with his smug artificial
smile, that was like a grimace in its best estate, and
now hardly seemed more than a contortion. But
beauty in any sense was not what the observer was
prepared to expect in Nehemiah, and the moonshiner seemed to accept the smile at its face value,
and to respect its intention.

"Spies don't kem climbin' down that thar path o'
yourn in full view through the water " — for the
landscape was as visible through the thin falling
sheet as if it had been the slightly corrugated glass

of a window—"do they?" Yerby asked, with a jocose intonation. "That thar shootin'-iron o' yourn liked ter hev skeered me ter death whenst I fust seen it."

His interlocutor pondered on this answer for a moment. He had an adviser among his corps whose opinion he evidently valued ; he exchanged a quick glance with one of the men who was but dimly visible in the shadows beyond the still, where there seemed to be a series of troughs leading a rill of running water down from some farther spring and through the tub in which the spiral worm was coiled. This man had a keen, white, lean face, with an ascetic, abstemious expression, and he looked less like a distiller than some sort of divine—some rustic pietist, with strange theories and unhappy speculations and unsettled mind. It was a face of subtle influences, and the very sight of it roused in Nehemiah a more heedful fear than the "shootin'-iron" in the bluff moonshiner's hand had induced. He was silent, while the other resumed the office of spokesman.

"Ye ain't 'quainted hyar"—he waved his hand with the pistol in it around at the circle of uncowering men, although the mere movement made Nehemiah cringe with the thought that an accidental discharge might as effectually settle his case as premeditated and deliberate murder. "Ye dun'no' none o' us. What air ye a-doin' hyar?"

"Why, that thar war the very trouble," Yerby hastily explained. "*I didn't know none o' ye!* I hed hearn ez thar war a still somewhars on Hide-an'-Seek Creek"—once more there ensued a swift exchange of glances among the party—"but nobody knew who run it nor whar 'twar. An' one day, consider'ble time ago, I war a-passin' nigh 'bouts an' I hearn that fiddle, an' that revealed the spot ter me. An' I kem ter-day 'lowin' ye an' me could strike a trade."

Once more the bluff man of force turned an anxious look of inquiry to the pale, thoughtful face in the brown and dark green shadows beyond the copper gleam of the still. If policy had required that Nehemiah should be despatched, his was the hand to do the deed, and his the stomach to support his conscience afterward. But his brain revolted from the discriminating analysis of Nehemiah's discourse and a decision on its merits.

" Trade fur what ?" he demanded at last, on his own responsibility, for no aid had radiated from the face which his looks had interrogated.

" Fur whiskey, o' course." Nehemiah made the final plunge boldly. " I be goin' ter open a store at the cross-roads, an' I 'lowed I could git cheaper whiskey untaxed than taxed. I 'lowed ye wouldn't make it ef ye didn't expec' ter sell it. I didn't know none o' you-uns, an' none o' yer customers. An' ez I expec' ter git mo' profit on sellin' whiskey 'n ennything else in the store, I jes took foot in hand an' kem ter see 'boutn it mysef. I never 'lowed, though, ez it mought look cur'ous ter you-uns, or like a spy, ter kem ez bold ez brass down the path in full sight."

The logic of the seeming security of his ap-. proach, and the apparent value of his scheme, had their full weight. He saw credulity gradually overpowering doubt and distrust, and his heart grew light with relief. Even their cautious demur, intimating a reserve of opinion to the effect that they would think about it, did not daunt him now. He believed, in the simplicity of his faith in his own craft, now once more in the ascendant, that if they should accept his proposition he would be free to go without further complication of his relations with wild-cat whiskey. He could not sufficiently applaud his wits for the happy termination of the adventure to which they had led him. He had gone no further in the matter than he had always intended. Brush whiskey was the commodity that addressed itself most to his sense of speculation. For this he had always expected to ferret out some way of safely negotiating. He had gone no further than he should have done, at all events, a little later. He even began mentally to "figger on the price" down to which he should be able to bring the distillers, as he accepted a proffered seat in the circle about the still. He could neither divide nor multiply by fractions, and it is not too much to say that he might have been throttled on the spot if the moonshiners could have had a mental vision of the liberties the stalwart integers were taking with their price-current, so to speak, and the preternatural discount that was making so free with their profits. So absorbed in this pleasing intellectual exercise was Nehemiah that he did not observe that any one had left the coterie; but when a stir without on the rocks intimated an approach he was suddenly ill at ease, and this discomfort increased when the new-comer proved to be a man who knew him.

"Waal, Nehemiah Yerby!" he exclaimed, shaking his friend's hand, "I never knowed you-uns ter be consarned in sech ez moonshinin'. I hev been a-neighborin' Isham hyar," he laid his heavy hand on the tall moonshiner's shoulder, "fur ten year an' better, but I won't hev nuthin' ter do with bresh whiskey or aidin' or abettin' in illicit 'stillin'. I like Isham, an' Isham he likes me, an' we hev jes agreed ter disagree."

Nehemiah dared not protest nor seek to explain. He could invent no story that would not give the lie direct to his representations to the moonshiners. He felt that their eyes were upon him. He could only hope that his silence did not seem to them like denial—and yet was not tantamount to confession in the esteem of his upbraider.

"Yes, sir," his interlocutor continued, "it's a mighty bad government ter run agin." Then he turned to the moonshiner, evidently taking up the business that had brought him here. "Lemme see what sorter brand ye hev registered fur yer cattle, Isham."

Yerby's heart sank when the suspicion percolated through his brain that this man had been induced to come here for the purpose of recognizing him. More fixed in this opinion was he when no description of the brand of the cattle could be found, and the visitor finally went away, his errand bootless.

From time to time during the afternoon other men went out and returned with recruits on various pretexts, all of which Nehemiah believed masked the marshalling of witnesses to incriminate him as one of themselves, in order to better secure his constancy to the common interests, and in case he was playing false to put others into possession of the facts as to the identity of the informer. His liability to the law for aiding and abetting in moonshining was very complete before the day darkened, and his jeopardy as to the information he had given made him shake in his shoes.

For at any moment, he reflected, in despair, the laggard raiders might swoop down upon them, and the choice of rôles offered to him was to seem to them a moonshiner, or to the moonshiners an informer. The first was far the safer, for the clutches of the law were indeed feeble as contrasted with the popular fury that would pursue him unwearied for years until its vengeance was accomplished. From the one, escape was to the last degree improbable; from the other, impossible.

Any pretext to seek to quit the place before the definite arrangements of his negotiation were consummated seemed even to him, despite his eagerness to be off, too tenuous, too transparent, to be essayed, although he devised several as he sat meditative and silent amongst the group about the still. The prospect grew less and less inviting as the lingering day waned, and the evening shadows, dank and chill, perceptibly approached. The brown and green recesses of the grotto were at once murkier, and yet more distinctly visible, for the glow of the fire, flickering through the crevices of the metal door of the furnace, had begun to assert its luminous quality, which was hardly perceptible in the full light of day, and brought out the depth of the shadows. The figures and faces of the moonshiners showed against the deepening gloom. The sunset

clouds were still red without ; a vague roseate suffusion was visible through the falling water. The sun
itself had not yet sunk, for an oblique and almost
level ray, piercing the cataract, painted a series of
faint prismatic tints on one side of the rugged arch.
But while the outer world was still in touch with
the clear-eyed day, night was presently here, with
mystery and doubt and dark presage. The voice
of Hoho-hebee Falls seemed to him louder, full of
strange, uncomprehended meanings, and insistent
iteration. Vague echoes were elicited. Sometimes
in a seeming pause he could catch their lisping
sibilant tones repeating, repeating—what ? As the
darkness encroached yet more heavily upon the
cataract, the sense of its unseen motion so close at
hand oppressed his very soul ; it gave an idea of
the swift gathering of shifting invisible multitudes,
coming and going — who could say whence or
whither ? So did this impression master his nerves
that he was glad indeed when the furnace door was
opened for fuel, and he could see only the inanimate, ever-descending sheet of water—the reverse
interior aspect of Hoho-hebee Falls—all suffused
with the uncanny tawny light, but showing white
and green tints like its diurnal outer aspect, instead
of the colorless outlines, resembling a drawing of a
cataract, which the cave knew by day. He did
not pause to wonder whether the sudden transient
illumination was visible without, or how it might
mystify the untutored denizens of the woods, bear,
or deer, or wolf, perceiving it aglow in the midst of
the waters like a great topaz, and anon lost in the
gloom. He pined to see it ; the momentary cessa-

tion of darkness, of the effect of the sounds, so
strange in the obscurity, and of the chill, pervasive
mystery of the invisible, was so grateful that its influence was tonic to his nerves, and he came to
watch for its occasion and to welcome it.   He did
not grudge it even when it gave the opportunity
for a close, unfriendly, calculating scrutiny of his
face by the latest comer to the still.   This was the
neighboring miller, also liable to the revenue laws,
the distillers being valued patrons of the mill, and
since he ground the corn for the mash he thereby
aided and abetted in the illicit manufacture of the
whiskey.   His life was more out in the world than
that of his underground *confrères*, and perhaps, as
he had a thriving legitimate business, and did not
live by brush whiskey, he had more to lose by de-
tection than they, and deprecated even more any
unnecessary risk.   He evidently took great um-
brage at the introduction of Nehemiah amongst
them.

   "Oh yes," he observed, in response to the cordial
greeting which he met ; "an' I'm glad ter see ye all
too.   I'm powerful glad ter kem ter the still enny
time.   It's ekal ter goin' ter the settlemint, or plumb
ter town on a County Court day.   Ye see *everybody*, an' hear *all* the news, an' meet up with *inter-
estin' strangers*.   I tell ye, now, the mill's plumb
lonesome compared ter the still, an' the mill's always hed the name of a place whar a heap o'
cronies gathered ter swap lies, an' sech."

   The irony of this description of the social delights
and hospitable accessibilities of a place esteemed
the very stronghold of secrecy itself—the liberty of

every man in it jeopardized by the slightest lapse
of vigilance or judgment—was very readily to be
appreciated by the group, who were invited by this
fair show of words to look down the vista of the
future to possible years of captivity in the jails of
far-away States as Federal prisoners. The men
gazed heavily and anxiously from one to another
as the visitor sank down on the rocks in a relaxed
attitude, his elbow on a higher ledge behind him,
supporting his head on his hand; his other hand
was on his hip, his arm stiffly akimbo, while he
looked with an expression of lowering exasperation
at Yerby. It was impossible to distinguish the col-
or of his garb, so dusted with flour was he from
head to foot; but his long boots drawn over his
trousers to the knee, and his great spurs, and a
brace of pistols in his belt, seemed incongruous ac-
cessories to the habiliments of a miller. His large,
dark hat was thrust far back on his head; his hair,
rising straight in a sort of elastic wave from his
brow, was powdered white; the effect of his florid
color and his dark eyes was accented by the contrast; his pointed beard revealed its natural tints
because of his habit of frequently brushing his hand
over it, and was distinctly red. He was lithe and
lean and nervous, and had the impatient temper
characteristic of mercurial natures. It mattered not
to him what was the coercion of the circumstances
which had led to the reception of the stranger here,
nor what was the will of the majority; he disap-
proved of the step; he feared it; he esteemed it a
grievance done him in his absence; and he could
not conceal his feelings nor wait a more fitting time

to express them in private. His irritation and objection evidently caused some solicitude amongst the others. He was important to them, and they deprecated his displeasure. Isham Beaton listened to the half-covert sneers of his words with perturbation plainly depicted on his face, and the man whom Nehemiah had at first noticed as one whose character seemed that of adviser, and whose opinion was valued, now spoke for the first time. He handed over a broken-nosed pitcher with the remark, "Try the flavor of this hyar whiskey, Alfred; 'pears like ter me the bes' we-uns hev ever hed."

His voice was singularly smooth; it had all the qualities of culture; every syllable, every lapse of his rude dialect, was as distinct as if he had been taught to speak in this way; his tones were low and even, and modulated to suave cadences; the ear experienced a sense of relief after the loud, strident voice of the miller, poignantly penetrating and pitched high.

"Naw, Hilary, I don't want nuthin' ter drink. 'Bleeged ter ye, but I ain't wantin' nuthin' ter drink," reiterated the miller, plaintively.

Isham Beaton cast a glance of alarm at the dimly seen, monastic face of his adviser in the gloom. It was unchanged. Its pallor and its keen outline enabled its expression to be discerned as he himself went through the motions of sampling the rejected liquor, shook his head discerningly, wiped his mouth on the back of his hand, and deposited the pitcher near by on a shelf of the rock.

A pause ensued. Nehemiah, with every desire

17

to be agreeable, hardly knew how to commend himself to the irate miller, who would have none of his very existence. No one could more eagerly desire him to be away than he himself. But his absence would not satisfy the miller; nothing less than that the intruder should never have been here. Every perceptible lapse of the moonshiners into anxiety, every recurrent intimation of their most pertinent reason for this anxiety, set Nehemiah a-shaking in his shoes. Should it be esteemed the greatest good to the greatest number to make safe-ly away with him, his fate would forever remain unknown, so cautious had he been to leave no trace by which he might be followed. He gazed with deprecating urbanity, and with his lips dis-tended into a propitiating smile, at the troubled face powdered so white and with its lowering eyes so dark and petulant. He noted that the small-talk amongst the others, mere unindividualized lump-ish fellows with scant voice in the government of their common enterprise, had ceased, and that they no longer busied themselves with the necessary work about the still, nor with the snickering inter-ludes and horse-play with which they were wont to beguile their labors. They had all seated them-selves, and were looking from one to the other of the more important members of the guild with an air which betokened the momentary expectation of a crisis. The only exception was the man who had the violin; with the persistent, untimely industry of incapacity, he twanged the strings, and tuned and retuned the instrument, each time producing a result more astonishingly off the key than before.

He was evidently unaware of this till some one
with senses ajar would suggest that all was not as
it should be in the drunken reeling catch he sought
to play, when he would desist in surprise, and once
more diligently rub the bow with rosin, as if that
mended the matter. The miller's lowering eyes rested on his shadowy outline as he sat thus engaged,
for a moment, and then he broke out suddenly :

"Yes, this hyar still is the place fur news, an'
the place ter look out fur what ye don't expec' ter
happen. It's powerful pleasant ter be a-meetin' of
folks hyar—this hyar stranger this evenin' "—his
gleaming teeth in the semi-obscurity notified Yerby
that a smile of spurious politeness was bent upon
him, and he made haste to grin very widely in response—"an' that thar fiddle 'minds me o' how
onexpected 'twar whenst I met up with Lee-yander
hyar—'pears ter me, Bob, ez ye air goin' ter diddle
the life out'n his fiddle—an' Hilary jes begged an'
beseeched me ter take the boy with me ter help
'round the mill, ez he war a-runnin' away. Ye
want me ter 'commodate this stranger too, ez
mebbe air runnin' from them ez wants him, hey
Hilary ?"

The grin was petrified on Nehemiah's face. He
felt his blood rush quickly to his head in the excitement of the moment. So here was the bird
very close at hand ! And here was his enterprise
complete and successful. He could go away after
the cowardly caution of the moonshiners should
have expended itself in dallying and delay, with
his negotiation for the "wild-cat" ended, and his
accomplished young relative in charge. He drew

himself erect with a sense of power. The moonshiners, the miller, would not dare to make an objection. He knew too much! he knew far too much!

The door of the furnace was suddenly flung ajar, but he was too much absorbed to perceive the change that came upon the keen face of Hilary Tarbetts, who knelt beside it, as the guest's portentous triumphant smile was fully revealed. Yerby did not lose, however, the glance of reproach which the moonshiner cast upon the miller, nor the miller's air at once triumphant, ashamed, and regretful. He had in petulant pique disclosed the circumstance which he had pledged himself not to disclose.

"This man's name is Yerby too," Hilary said, significantly, gazing steadily at the miller.

The miller looked dumfounded for a moment. He stared from one to the other in silence. His conscious expression changed to obvious discomfiture. He had expected no such result as this. He had merely given way to a momentary spite in the disclosure, thinking it entirely insignificant, only calculated to slightly annoy Hilary, who had made the affair his own. He would not in any essential have thwarted his comrade's plans intentionally, nor in his habitual adherence to the principles of fair play would he have assisted in the boy's capture. He drew himself up from his relaxed posture; his spurred feet shuffled heavily on the stone floor of the grotto. A bright red spot appeared on each cheek; his eyes had become anxious and subdued in the quick shiftings of temper common to the red-haired gentry; his face of helpless appeal was bent on Hilary Tarbetts, as if relying on his resources to mend the matter; but ever and anon he turned his eyes, animated with a suspicious dislike, on Yerby, who, however, could have snapped his fingers in the faces of them all, so confident, so hilariously triumphant was he.

"Yerby, I b'lieve ye said yer name war, an' so did Peter Green," said Tarbetts, still kneeling by the open furnace door, his pale cheek reddening in the glow of the fire.

Thus reminded of the testimony of his acquaintance, Yerby did not venture to repudiate his cognomen.

"An' what did ye kem hyar fur?" blustered the miller. "A-sarchin' fur the boy?"

Yerby's lips had parted to acknowledge this fact, but Tarbetts suddenly anticipated his response, and answered for him:

"Oh no, Alfred. Nobody ain't sech a fool ez ter kem hyar ter this hyar still, a stranger an' mebbe suspected ez a spy, ter hunt up stray children, an' git thar heads shot off, or mebbe drownded in a mighty handy water-fall, or sech. This hyar man air one o' we-uns. He air a-tradin' fur our liquor, an' he'll kerry a barrel away whenst he goes."

Yerby winced at the suggestion conveyed so definitely in this crafty speech; he was glad when the door of the furnace closed, so that his face might not tell too much of the shifting thoughts and fears that possessed him.

The miller's fickle mind wavered once more. If Yerby had not come for the boy, he himself had

done no damage in disclosing Leander's whereabouts. Once more his quickly illumined anger
was kindled against Tarbetts, who had caused him
a passing but poignant self-reproach. " Waal, then,
Hilary," he demanded, "what air ye a-raisin' sech
a row fur? Lee-yander ain't noways so special precious ez I knows on. Toler'ble lazy an' triflin', an'
mightily gi'n over ter ˙moonin' over a readin'-book
he hev got. That thar mill war a-grindin' o' nuthin'
at all more'n haffen ter-day, through me bein' a-nap-
pin', and Lee-yander plumb demented by his book so
ez he furgot ter pour enny grist inter the hopper.
Shucks! his kin is welcome ter enny sech critter ez
that, though I ain't denyin' ez he'd be toler'ble spry
ef he could keep his nose out'n his book," he qualified, relenting, "or his fiddle out'n his hands. I
made him leave his fiddle hyar ter the still, an' I be
goin' ter hide his book."

"No need," thought Nehemiah, scornfully. Book
and scholar and it might be fiddle too, so indulgent
had the prospect of success made him, would by to-
morrow be on the return route to the cross-roads.
He even ventured to differ with the overbearing
miller.

"I dun'no' 'bout that; books an' edication in
gin'ral air toler'ble useful wunst in a while;" he was
thinking of the dark art of dividing and multiplying
by fractions. "The Yerbys hev always hed the
name o' bein' quick at thar book."

Now the democratic sentiment in this country is
bred in the bone, and few of its denizens have so
diluted it with Christian grace as to willingly acknowledge a superior. In such a coterie as this

"eating humble-pie" is done only at the muzzle of a "shootin'-iron."

" Never hearn afore ez enny o' the Yerbys knowed B from bull-foot," remarked one of the unindividualized lumpish moonshiners, shadowy, indistinguishable in the circle about the rotund figure of the still. He yet retained acrid recollections of unavailing struggles with the alphabet, and was secretly of the opinion that education was a painful thing, and, like the yellow - fever or other deadly disease, not worth having. Nevertheless, since it was valued by others, the Yerbys should scathless make no unfounded claims. "Ef the truth war knowed, nare one of 'em afore could tell a book from a bear-trap."

Nehemiah's flush the darkness concealed ; he moistened his thin lips, and then gave a little cackling laugh, as if he regarded this as pleasantry. But the demolition of the literary pretensions of his family once begun went bravely on.

"Abner Sage larnt this hyar boy all he knows," another voice took up the testimony. " Ab 'lows ez his mother war quick at school, but his dad—law I I knowed Ebenezer Yerby I He war a frien'ly sorter cuss, good-nachured an' kind-spoken, but ye could put all the larnin' he hed in the corner o' yer eye."

"An' Lee-yander don't favor none o' ye," observed another of the undiscriminated, unimportant members of the group, who seemed to the groping scrutiny of Nehemiah to be only endowed with sufficient identity to do the rough work of the still, and to become liable to the Federal law. " Thar's

Hil'ry—he seen it right off. Hil'ry he tuk a look at Lee-yander whenst he wanted ter kem an' work along o' we-uns, 'kase his folks wanted ter take him away from the Sudleys. Hil'ry opened the furnace door—jes so ; an' he cotch the boy by the arm "— the great brawny fellow, unconsciously dramatic, suited the action to the word, his face and figure illumined by the sudden red glow—" an' Hil'ry, he say, 'Naw, by God—ye hev got yer mother's eyes in yer head, an' I'll swear ye sha'n't larn ter be a sot!' An' that's how kem Hil'ry made Alf Bixby take Lee-yander ter work in the mill. Ef ennybody tuk arter him he war convenient ter disappear down hyar with we-uns. So he went ter the mill."

"An' I wisht I hed put him in the hopper an' ground him up," said the miller, in a blood-curdling tone, but with a look of plaintive anxiety in his eyes. "He hev made a heap o' trouble 'twixt Hil'ry an' me fust an' last. Whar's Hil'ry disappeared to, en-nyways?"

For the flare from the furnace showed that this leading spirit amongst the moonshiners had gone softly out. Nehemiah, whose courage was dis-sipated by some subtle influence of his presence, now made bold to ask, " An' what made him ter set store on Lee-yander's mother's eyes?" His tone was as bluffly sarcastic as he dared.

" Shucks—ye mus' hev hearn that old tale," said the miller, cavalierly. "This hyar Malviny Hixon —ez lived down in Tanglefoot Cove then—her an' Hil'ry war promised ter marry, but the revenuers captured him—he war a-runnin' a still in Tangle-foot then — an' they kep' him in jail somewhar in

the North fur five year. Waal, she waited toler'ble
constant fur two or three year, but Ebenezer Yerby
he kem a-visitin' his kin down in Tanglefoot Cove,
an' she an' him met at a bran dance, an' the fust
thing I hearn they war married, an' 'fore Hil'ry got
back she war dead an' buried, an' so war Ebenezer."

There was a pause while the flames roared in the
furnace, and the falling water desperately dashed
upon the rocks, and its tumultuous voice continuously pervaded the silent void wildernesses without,
and the sibilant undertone, the lisping whisperings,
smote the senses anew.

"He met up with cornsider'ble changes fur five
year," remarked one of the men, regarding the matter in its chronological aspect.

Nehemiah said nothing. He had heard the story
before, but it had been forgotten. A worldly mind
like his is not apt to burden itself with the sentimental details of an antenuptial romance of the
woman whom his half-brother had married many
years ago.

A persuasion that it was somewhat unduly long-
lived impressed others of the party.

"It's plumb cur'us Hil'ry ain't never furgot her,"
observed one of them. "He hev never married at
all. My wife says it's jes contrariousness. Ef Malviny hed been his wife an' died, he'd hev married
agin 'fore the year war out. An' I tell my wife
that he'd hev been better acquainted with her then,
an' would hev fund out ez no woman war wuth
mournin' 'bout fur nigh twenty year. My wife says
she can't make out ez how Hil'ry 'ain't got pride

enough not ter furgive her fur givin' him the mitten like she done. An' I tell my wife that holdin' a gredge agin a woman fur bein' fickle is like holdin' a gredge agin her fur bein' a woman."

He paused with an air, perceived somehow in the brown dusk, of having made a very neat point. A stir of assent was vaguely suggested when some chivalric impulse roused a champion at the farther side of the worm, whose voice rang out brusquely:

"Jes listen at Tom! A body ter hear them tales he tells 'bout argufyin' with his wife would 'low he war a mighty smart, apt man, an' the pore foolish 'oman skeercely hed a sensible word ter bless her-self with. When everybody that knows Tom knows he sings mighty small round home. Ye stopped too soon, Tom. Tell what yer wife said to that."

Tom's embarrassed feet shuffled heavily on the rocks, apparently in search of subterfuge. The dazzling glintings from the crevices of the furnace door showed here and there gleaming teeth broadly agrin.

"Jes called me a fool in gineral," admitted the man skilled in argument.

"An' didn't she 'low ez men folks war fickle too, an' remind ye o' yer young days whenst ye went a-courtin' hyar an' thar, an' tell over a string o' gals' names till she sounded like an off'cer callin' the roll?"

"Ye-es," admitted Tom, thrown off his balance by this preternatural insight, "but all them gals war a-tryin' ter marry me — not me tryin' ter marry them."

There was a guffaw at this modest assertion, but

the disaffected miller's tones dominated the rude merriment.

"Whenst a feller takes ter drink folks kin spell out a heap o' reasons but the true one—an' that's 'kase he likes it. Hil'ry 'ain't never named that 'oman's name ter me, an' I hev knowed him ez well ez ennybody hyar. Jes t'other day whenst that boy kem, bein' foolish an' maudlin, he seen suthin' on-common in Lee-yander's eyes — they'll be mighty oncommon ef he keeps on readin' his tomfool book, ez he knows by heart, by the firelight when it's dim. Ef folks air so sot agin strong drink, let 'em drink less tharsefs. Hear Brother Peter Vickers preach agin liquor, an' ye'd know ez all wine - bibbers air bound fur hell."

"But the Bible don't name 'whiskey' once," said the man called Tom, in an argumentative tone. "Low wines I'll gin ye up;" he made the discrim-ination in accents betokening much reasonable ad-mission; "but nare time does the Bible name whis-key, nor yit peach brandy, nor apple-jack."

"Nor cider nor beer," put in an unexpected re-cruit from the darkness.

The miller was silent for a moment, and gave token of succumbing to this unexpected polemic strength. Then, taking thought and courage to-gether, "Ye can't say the Bible ain't down on 'strong drink'?" There was no answer from the vanquished, and he went on in the overwhelming miller's voice: "Hil'ry hed better be purtectin' his-self from strong drink, 'stiddier the boy—by makin' him stay up thar at the mill whar he knows thar's no drinkin' goin' on—ez will git chances at it other

ways, ef not through him, in the long life he hev got
ter live. The las' time the revenuers got Hil'ry
'twar through bein' ez drunk ez a fraish-biled owl.
It makes me powerful oneasy whenever I know ye
air all drunk an' a-gallopadin' down hyar, an' no mo'
able to act reasonable in case o' need an' purtect
yersefs agin spies an' revenuers an' sech 'n nuthin'
in this worl'. The las' raid, ye 'member, we hed the
still over yander;" he jerked his thumb in the direction present to his thoughts, but unseen by his coad-
jutors; "a man war wounded, an' we dun'no' but
what killed in the scuffle, an' it mought be a hangin' matter ter git caught now. Ye oughter keep
sober; an' ye know, Isham, ye oughter keep Hil'ry
sober. I dun'no' why ye can't. I never could
abide the nasty stuff — it's enough ter turn a bull-
frog's stomach. Whiskey is good ter sell—not ter
drink. Let them consarned idjits in the flat woods
buy it, an' drink it. Whiskey is good ter sell—not
ter drink."

This peculiar temperance argument was received
in thoughtful silence, the reason of all the mountaineers commending it, while certain of them knew
themselves and were known to be incapable of prof-
iting by it.

Nehemiah had scant interest in this conversation.
He was conscious of the strain on his attention as
he followed it, that every point of the situation
should be noted, and its utility canvassed at a leisure
moment. He marked the allusion to the man supposed to have been killed in the skirmish with the
raiders, and he appraised its value as coercion in
any altercation that he might have in seeking to

take Leander from his present guardians. But he felt in elation that this was likely to be of the slightest; the miller evidently found himself hampered rather than helped by the employment of the boy; and as to the moonshiner's sentimental partisanship, for the sake of an old attachment to the dead-and-gone mountain girl, there was hardly anything in the universe so tenuous as to bear comparison with its fragility. "A few drinks ahead," he said to himself, with a sneer, "an' he won't remember who Malviny Hixon was, ef thar is ennything in the old tale—which it's more'n apt thar ain't."

He began, after the fashion of successful people, to cavil because his success was not more complete. How the time was wasting here in this uncomfortable interlude! Why could he not have discovered Leander's whereabouts earlier, and by now be jogging along the road home with the boy by his side? Why had he not bethought himself of the mill in the first instance—that focus of gossip where all the news of the countryside is mysteriously garnered and thence dispensed bounteously to all comers? It was useless, as he fretted and chafed at these untoward omissions, to urge in his own behalf that he did not know of the existence of the mill, and that the miller, being an ungenial and choleric man, might have perversely lent himself to resisting his demand for the custody of the young runaway. No, he told himself emphatically, and with good logic, too, the miller's acrimony rose from the fact of a stranger's discovery of the still and the danger of his introduction into its charmed circle. And that reflection reminded him anew of his own danger here

—not from the lawless denizens of the place, but from the forces which he himself had evoked, and again he glanced out toward the water-fall as fearful of the raiders as any moonshiner of them all.

But what sudden glory was on the waters, mystic, white, an opaque brilliance upon the swirling foam and the bounding spray, a crystalline glitter upon the smooth expanse of the swift cataract! The moon was in the sky, and its light, with noiseless tread, sought out strange, lonely places, and illusions were astir in the solitudes. Pensive peace, thoughts too subtle for speech to shape, spiritual yearnings, were familiars of the hour and of this melancholy splendor; but he knew none of them, and the sight gave him no joy. He only thought that this was a night for the saddle, for the quiet invasion of the woods, when the few dwellers by the way-side were lost in slumber. He trembled anew at the thought of the raiders whom he himself had summoned; he forgot his curses on their laggard service; he upbraided himself again that he had not earlier made shift to depart by some means—by any means—before the night came with this great emblazoning bold-faced moon that but prolonged the day; and he started to his feet with a galvanic jerk and a sharp exclamation when swift steps were heard on the rocks outside, and a man with the lightness of a deer sprang down the ledges and into the great arched opening of the place.

"'Tain't nobody but Hil'ry," observed Isham Beaton, half in reproach, half in reassurance.

The pervasive light without dissipated in some degree the gloom within the grotto; a sort of gray

visibility was on the appurtenances and the figures
about the still, not strong enough to suggest color,
but giving contour. His fright had been marked,
he knew; a sort of surprised reflectiveness was in
the manner of several of the moonshiners, and Nehemiah, with his ready fears, fancied that this inopportune show of terror had revived their suspicions
of him. It required some effort to steady his nerves
after this, and when footfalls were again audible outside, and all the denizens of the place sat calmly
smoking their pipes without so much as a movement
toward investigating the sound, he, knowing whose
steps he had invited thither, had great ado with the
coward within to keep still, as if he had no more reason to fear an approach than they.

A great jargon in the tone of ecstasy broke suddenly on the air upon this new entrance, shattering what little composure Nehemiah had been able
to muster; a wide-mouthed exaggeration of welcome
in superlative phrases and ready chorus. Swiftly
turning, he saw nothing for a moment, for he looked
at the height which a man's head might reach, and
the new-comer measured hardly two feet in stature,
waddled with a very uncertain gait, and although he
bore himself with manifest complacence, he had evidently heard the like before, as he was jovially hailed
by every ingratiating epithet presumed to be accept-
able to his infant mind. He was attended by a tall,
gaunt boy of fifteen, barefooted, with snaggled teeth
and a shock of tow hair, wearing a shirt of unbleached
cotton, and a pair of trousers supported by a single
suspender drawn across a sharp, protuberant shoul-
der-blade behind and a very narrow chest in front.

But his face was proud and happy and gleeful, as if he occupied some post of honor and worldly emolument in attending upon the waddling wonder on the floor in front of him, instead of being assigned the ungrateful task of seeing to it that a very ugly baby closely related to him did not, with the wiliness and ingenuity of infant nature, invent some method of making away with himself. For he *was* an ugly baby as he stood revealed in the flare of the furnace door, thrown open that his admirers and friends might feast their eyes upon him. His short wisps of red hair stood straight up in front; his cheeks were puffy and round, but very rosy; his eyes were small and dark, but blandly roguish; his mouth was wide and damp, and had in it a small selection of sample teeth, as it were; he wore a blue checked homespun dress garnished down the back with big horn buttons, sparsely set on; he clasped his chubby hands upon a somewhat pompous stomach; he sidled first to the right, then to the left, in doubt as to which of the various invitations he should accept.

"Kem hyar, Snooks!" "Right hyar, Toodles!" "Me hyar, Monkey Doodle!" "Hurrah fur the leetle-est moonshiner on record!" resounded fulsomely about him. Many were the compliments showered upon him, and if his flatterers told lies, they had told more wicked ones. The pipes all went out, and the broken-nosed pitcher languished in disuse as he trotted from one pair of outstretched arms to another to give an exhibition of his progress in the noble art of locomotion; and if he now and again sat down, unexpectedly to himself and to the spectator, he was promptly put upon his feet again with spurious applause and encouragement. He gave an exhibition of his dancing—a funny little shuffle of exceeding temerity, considering the facilities at his command for that agile amusement, but he was made reckless by praise—and they all lied valiantly in chorus. He repeated all the words he knew, which were few, and for the most part unintelligible, crowed like a cock, barked like a dog, mewed like a cat, and finally went away, his red cheeks yet more ruddily aglow, grave and excited and with quickly beating pulses, like one who has achieved some great public success and led captive the hearts of thousands.

The turmoils of his visit and his departure were great indeed. It all irked Nehemiah Yerby, who had scant toleration of infancy and little perception of the jocosity of the aspect of callow human nature, and it seemed strange to him that these men, all with their liberty, even their existence, jeopardized upon the chances that a moment might bring forth, could so relax their sense of danger, so disregard the mandates of stolid common-sense, and give themselves over to the puerile beguilements of the visitor. The little animal was the son of one of them, he knew, but he hardly guessed whom until he marked the paternal pride and content that had made unwontedly placid the brow of the irate miller while the ovation was in progress. Nehemiah greatly preferred the adult specimen of the race, and looked upon youth as an infirmity which would mend only with time. He was easily confused by a stir; the gurglings, the ticklings, the loud laughter both in the deep bass of the hosts and the keen treble of the guest had a befuddling effect upon him; his

18

powers of observation were numbed. As the great, burly forms shifted to and fro, resuming their former places, the red light from the open door of the furnace illumining their laughing, bearded countenances, casting a roseate suffusion upon the white turmoils of the cataract, and showing the rugged interior of the place with its damp and dripping ledges, he saw for the first time among them Leander's slight figure and smiling face ; the violin was in his hand, one end resting on a rock as he tightened a string ; his eyes were bent upon the instrument, while his every motion was earnestly watched by the would-be fiddler.

Nehemiah started hastily to his feet. He had not expected that the boy would see him here. To share with one of his own household a secret like this of aiding in illicit distilling was more than his hardihood could well contemplate. As once more the contemned " ping-pang " of the process of tuning fell upon the air, Leander chanced to lift his eyes. They smilingly swept the circle until they rested upon his uncle. They suddenly dilated with astonishment, and the violin fell from his nerveless hand upon the floor. The surprise, the fear, the repulsion his face expressed suddenly emboldened Nehemiah. The boy evidently had not been prepared for the encounter with his relative here. Its only significance to his mind was the imminence of capture and of being constrained to accompany his uncle home. He cast a glance of indignant reproach upon Hilary Tarbetts, who was not even looking at him. The moonshiner stood filling his pipe with tobacco, and as he deftly extracted a coal

from the furnace to set it alight, he shut the door
with a clash, and for a moment the whole place
sunk into invisibility, the vague radiance vouchsafed
to the recesses of the grotto by the moonbeams on
the water without annihilated for the time by the
contrast with the red furnace glare. Nehemiah had
a swift fear that in this sudden eclipse Leander
might slip softly out and thus be again lost to him,
but as the dull gray light gradually reasserted itself,
and the figures and surroundings emerged from the
gloom, resuming shape and consistency, he saw
Leander still standing where he had disappeared
in the darkness ; he could even distinguish his pale
face and lustrous eyes. Leander at least had no
intention to shirk explanations.

"Why, Uncle Nehemiah!" he said, his boyish
voice ringing out tense and excited above the tones
of the men, once more absorbed in their wonted interests. A sudden silence ensued amongst them.
"What air ye a-doin' hyar?"

"Waal, ah, Lee-yander, boy—" Nehemiah hesitated. A half-suppressed chuckle among the men,
whom he had observed to be addicted to horse-play,
attested their relish of the situation. Ridicule is
always of unfriendly intimations, and the sound
served to put Nehemiah on his guard anew. He
noticed that the glow in Hilary's pipe was still and
dull : the smoker did not even draw his breath as
he looked and listened. Yerby did not dare avow
the true purpose of his presence after his representations to the moonshiners, and yet he could not,
he would not in set phrase align himself with the
illicit vocation. The boy was too young, too irre-

sponsible, too inimical to his uncle, he reflected in
a sudden panic, to be intrusted with this secret. If
in his hap-hazard, callow folly he should turn informer, he was almost too young to be amenable to
the popular sense of justice. He might, too, by
some accident rather than intention, divulge the
important knowledge so unsuitable to his years and
his capacity for guarding it. He began to share
the miller's aversion to the introduction of outsiders
to the still. He felt a glow of indignation, as if he
had always been a party in interest, that the common safety should not be more jealously guarded.
The danger which Leander's youth and inexperience
threatened had not been so apparent to him when
he first heard that the boy had been here, and the
menace was merely for the others. As he felt the
young fellow's eyes upon him he recalled the ef-
fusive piety of his conversation at Tyler Sudley's
house, his animadversions on violin - playing and
liquor - drinking, and Brother Peter Vickers's mild
and merciful attitude toward sinners in those un-
spiced sermons of his, that held out such affluence
of hope to the repentant rather than to the self-
righteous. The blood surged unseen into Nehemiah's face. For shame, for very shame he could
not confess himself one with these outcasts. He
made a feint of searching in the semi-obscurity for
the rickety chair on which he had been seated, and
resumed his former attitude as Leander's voice once
more rang out:

"What air ye a-doin' hyar, Uncle Nehemiah?"

"Jes a-visitin', sonny; jes a-visitin'."

There was a momentary pause, and the felicity of

the answer was demonstrated by another chuckle
from the group. His senses, alert to the emergency,
discriminated a difference in the tone. This time
the laugh was with him rather than at him. He
noted, too, Leander's dumfounded pause, and the
suggestion of discomfiture in the boy's lustrous eyes,
still widely fixed upon him. As Leander stooped to
pick up the violin he remarked with an incidental
accent, and evidently in default of retort, "I be
powerful s'prised ter view ye hyar."

Nehemiah smarted under the sense of unmerited
reproach; so definitely aware was he of being out
of the character which he had assumed and worn
until it seemed even to him his own, that he felt as
if he were constrained to some ghastly masquerade.
Even the society of the moonshiners as their guest
was a reproach to one who had always piously, and
in such involuted and redundant verbiage, spurned
the ways and haunts of the evil-doer. According to
the dictates of policy he should have rested content
with his advantage over the silenced lad. But his
sense of injury engendered a desire of reprisal, and
he impulsively carried the war into the enemy's
country.

"I ain't in no ways s'prised ter view you-uns hyar,
Lee-yander," he said. "From the ways, Lee-yander,
ez ye hev been brung up by them slack-twisted Sudleys—ungodly folks 'ceptin' what little regeneration
they kin git from the sermons of Brother Peter Vickers, who air onsartin in his mind whether folks ez
ain't church-members air goin' ter be damned or no
—I ain't s'prised none ter view ye hyar." He suddenly remembered poor Laurelia's arrogations of

special piety, and it was with exceeding ill will that he added : "An' Mis' Sudley in partic'lar. Ty ain't no great shakes ez a shoutin' Christian. I dun'no' ez I ever hearn him shout once, but his wife air one o' the reg'lar, mournful, unrejicing members, always questioning the decrees of Providence, an' what ain't no nigher salvation, ef the truth war knowed, 'n a sinner with the throne o' grace yit ter find."

Leander had not picked up the violin ; this dis-quisition had arrested his hand until his intention was forgotten. He came slowly to the perpendicu-lar, and his eyes gleamed in the dusk. A vibration of anger was in his voice as he retorted :

"Mebbe so—mebbe they air sinners ; but they'd look powerful comical 'visitin'' hyar !"

"Ty Sudley ain't one o' the drinkin' kind," inter-polated the miller, who evidently had the makings of a temperance man. "He never sot foot hyar in his life."

"Them ez kem a-visitin' hyar," blustered the boy, full of the significance of his observations and ex-perience, "air either wantin' a drink or two 'thout payin' fur it, or else air tradin' fur liquor ter sell, an' that's the same ez moonshinin' in the law."

There was a roar of delight from the circle of lumpish figures about the still which told the boy that he had hit very near to the mark. Nehemiah hardly waited for it to subside before he made an effort to divert Leander's attention.

"An' what air *you-uns* doin' hyar ?" he demanded. "Tit for tat."

"Why," bluffly declared Leander, "I be a-runnin' away from you-uns. An' I 'lowed the still war one place whar I'd be sure o' not meetin' ye. Not ez I hev got ennything agin moonshinin' nuther," he added, hastily, mindful of a seeming reflection on his refuge. "Moonshinin' *is business*, though the United States don't seem ter know it. But I hev hearn ye carry on so pious 'bout not lookin' on the wine whenst it be red, that I 'lowed ye wouldn't like ter look on the still whenst—whenst it's yaller." He pointed with a burst of callow merriment at the big copper vessel, and once more the easily excited mirth of the circle burst forth irrepressibly.

Encouraged by this applause, Leander resumed: "Why, *I* even turns my back on the still myself out'n respec' ter the family—Cap'n an' Neighbor bein' so set agin liquor. Cap'n's ekal ter preachin' on it ef ennything onexpected war ter happen ter Brother Vickers. An' .when I *hev* ter view it, I look at it sorter cross-eyed." The flickering line of light from the crevice of the furnace door showed that he was squinting frightfully, with the much-admired eyes his mother had bequeathed to him, at the rotund shadow, with the yellow gleams of the metal barely suggested in the brown dusk. "So I tuk ter workin' at the mill. An' *I* hev got nuthin' ter do with the still." There was a pause. Then, with a strained tone of appeal in his voice, for a future with Uncle Nehemiah had seemed very terrible to him, "So ye warn't a-sarchin' hyar fur me, war ye, Uncle Nehemiah?"

Nehemiah was at a loss. There is a peculiar glutinous quality in the resolve of a certain type of character which is not allied to steadfastness of purpose, nor has it the enlightened persistence of obstinacy. In view of his earlier account of his purpose he

could not avow his errand; it bereft him of naught
to disavow it, for Uncle Nehemiah was one of those
gifted people who, in common parlance, do not mind
what they say.   Yet his reluctance to assure Leander that he was not the quarry that had led him into
these wilds so mastered him, the spurious relinquishment had so the aspect of renunciation, that he hesitated, started to speak, again hesitated, so palpably
that Hilary Tarbetts felt impelled to take a hand in
the game.

"Why don't ye sati'fy the boy, Yerby?" he said,
brusquely.   He took his pipe out of his mouth and
turned to Leander.   "Naw, bub.   He's jes tradin' fur
bresh whiskey, that's all; he's sorter skeery 'bout
bein' a wild-catter, an' he didn't want ye ter know
it."

The point of red light, the glow of his pipe, the
only exponent of his presence in the dusky recess
where he sat, shifted with a quick, decisive motion
as he restored it to his lips.

The blood rushed to Nehemiah's head; he was
dizzy for a moment; he heard his heart thump heavily; he saw, or he fancied he saw, the luminous dis-
tention of Leander's eyes as this Goliath of his bat-
tles was thus delivered into his hands.   To meet
him here proved nothing; the law was not violated
by Nehemiah in the mere knowledge that illicit
whiskey was in process of manufacture; a dozen
different errands might have brought him.   But this
statement put a sword, as it were, into the boy's
hands, and he dared not deny it.

"'Pears ter me," he blurted out at last, "ez ye air
powerful slack with yer jaw."

"Lee-yander ain't," coolly returned Tarbetts. "He knows all thar is ter know 'bout we-uns — an' why air ye not ter share our per'ls?"

"I ain't likely ter tell," Leander jocosely reassured him. "But I can't help thinkin' how it would rejice that good Christian 'oman, Cap'n Sudley, ez war made ter set on sech a low stool 'bout my pore old fiddle."

And thus reminded of the instrument, he picked it up, and once more, with the bow held aloft in his hand, he dexterously twanged the strings, and with his deft fingers rapidly and discriminatingly turned the screws, this one up and that one down. The earnest would-be musician, who had languished while the discussion was in progress, now plucked up a freshened interest, and begged that the furnace door might be set ajar to enable him to watch the process of tuning and perchance to detect its subtle secret. No objection was made, for the still was nearly empty, and arrangements tending to replenishment were beginning to be inaugurated by several of the men, who were examining the mash in tubs in the further recesses of the place. They were lighted by a lantern which, swinging to and fro as they moved, sometimes so swiftly as to induce a temporary fluctuation threatening eclipse, suggested in the dusk the erratic orbit of an abnormally magnified fire-fly. It barely glimmered, the dullest point of white light, when the rich flare from the opening door of the furnace gushed forth and the whole rugged interior was illumined with its color. The inadequate moonlight fell away; the chastened white splendor on the foam of the cataract, the crystalline glitter, timorous-

ly and elusively shifting, were annihilated; the swiftly descending water showed from within only a continuously moving glow of yellow light, all the brighter from the dark-seeming background of the world
glimpsed without. A wind had risen, unfelt in these
recesses and on the weighty volume of the main
sheet of falling water, but at its verge the fitful
gusts diverted its downward course, tossing slender
jets aslant, and sending now and again a shower of
spray into the cavern. Nehemiah remembered his
rheumatism with a shiver. The shadows of the men,
instead of an unintelligible comminglement with the
dusk, were now sharp and distinct, and the light
grotesquely duplicated them till the cave seemed full
of beings who were not there a moment before—
strange gnomes, clumsy and burly, slow of movement, but swift and mysterious of appearance and
disappearance. The beetling ledges here and there
imprinted strong black similitudes of their jagged
contours on the floor; with the glowing, weird illumination the place seemed far more uncanny than before, and Leander, with his face pensive once more
in response to the gentle strains slowly elicited by
the bow trembling with responsive ecstasy, his large
eyes full of dreamy lights, his curling hair falling
about his cheek as it rested upon the violin, his figure, tall and slender and of an adolescent grace,
might have suggested to the imagination a reminiscence of Orpheus in Hades. They all listened in languid pleasure, without the effort to appraise the music or to compare it with other per-
formances—the bane of more cultured audiences;
only the ardent amateur, seated close at hand on a

bowlder, watched the bowing with a scrutiny which betokened earnest anxiety that no mechanical trick might elude him. The miller's half-grown son, whose ear for any fine distinctions in sound might be presumed to have been destroyed by the clamors of the mill, sat a trifle in the background, and sawed away on an imaginary violin with many flourishes and all the exaggerations of mimicry; he thus furnished the zest of burlesque relished by the devotees of horse-play and simple jests, and was altogether una-ware that he had a caricature in his shadow just be-hind him, and was doing double duty in making both Leander and himself ridiculous. Sometimes he paused in excess of interest when the music elicited an amusement more to his mind than the long-drawn, pathetic cadences which the violinist so much affected. For in sudden changes of mood and in effective contrast the tones came showering forth in keen, quick staccato, every one as round and distinct as a globule, but as unindividualized in the swift ex-uberance of the whole as a drop in a summer's rain; the bow was but a glancing line of light in its rapidity, and the bounding movement of the theme set many a foot astir marking time. At last one young fellow, an artist too in his way, laid aside his pipe and came out to dance. A queer *pas seul* it might have been esteemed, but he was light and agile and not un-graceful, and he danced with an air of elation—al-beit with a grave face—which added to the enjoy-ment of the spectator, for it seemed so slight an effort. He was long-winded, and was still bounding about in the double-shuffle and the pigeon-wing, his shadow on the wall nimbly following every motion, when the violin's cadence quavered off in a discordant wail, and Leander, the bow pointed at the waterfall, exclaimed: "Look out! Somebody's thar! Out thar on the rocks!"

It was upon the instant, with the evident intention of a surprise, that a dozen armed men rushed precipitately into the place. Nehemiah, his head awhirl, hardly distinguished the events as they were confusedly enacted before him. There were loud, excited calls, unintelligible, mouthing back in the turbulent echoes of the place, the repeated word "Surrender!" alone conveying meaning to his mind. The sharp, succinct note of a pistol-shot was a short answer. Some quick hand closed the door of the furnace and threw the place into protective' gloom. He was vaguely aware that a prolonged struggle that took place amongst a group of men near him was the effort of the intruders to reopen it. All unavailing. He presently saw figures drawing back to the doorway out of the *mêlée*, for moonshiner and raider were alike indistinguishable, and he became aware that both parties were equally desirous to gain the outer air. Once more pistol-shots—outside this time — then a tumult of frenzied voices. Struck by a pistol-ball, Tarbetts had fallen from the ledge under the weight of the cataract and into the deep abysses below. The raiders were swiftly getting to saddle again. Now and then a crack mountain shot drew a bead upon them from the bushes; but mists were gathering, the moon was uncertain, and the flickering beams deflected the aim. Two or three of the horses lay dead on the river-bank, and others carried double, ridden by men with rid-

dled hats. They were in full retreat, for the catas-
trophe on the ledge of the cliff struck dismay to
their hearts. Had the man been shot, according to
the expectation of those who resist arrest, this would
be merely the logical sequence of events. But to be
hurled from a crag into a cataract savored of atroc-
ity, and they dreaded the reprisals of capture.

It was soon over. The whole occurrence, charged
with all the definitiveness of fate, was scant ten minutes in transition. A laggard hoof-beat, a faint echo
amidst the silent gathering of the moonlit mists, and
the loud plaint of Hoho-hebee Falls were the only
sounds that caught Nehemiah's anxious ear when
he crept out from behind the empty barrels and
tremulously took his way along the solitary ledges,
ever and anon looking askance at his shadow, that
more than once startled him with a sense of unwelcome companionship. The mists, ever thickening,
received him into their midst. However threatening
to the retreat of the raiders, they were friendly to
him. Once, indeed, they parted, showing through
the gauzy involutions of their illumined folds the
pale moon high in the sky, and close at hand a
horse's head just above his own, with wild, dilated
eyes and quivering nostrils. Its effect was as detached as if it were only drawn upon a canvas; the
mists rolled over anew, and but that he heard the
subdued voice of the rider urging the animal on,
and the thud of the hoofs farther away, he might
have thought this straggler from the revenue party
some wild illusion born of his terrors.

The fate of Hilary Tarbetts remained a mystery.
When the stream was dragged for his body it was

deemed strange that it should not be found, since the bowlders that lay all adown the rocky gorge so interrupted the sweep of the current that so heavy, a weight seemed likely to be caught amongst them. Others commented on the strength and great momentum of the flow, and for this reason it was thought that in some dark underground channel of Hide-and-Seek Creek the moonshiner had found his sepulchre. A story of his capture was circulated after a time ; it was supposed that he dived and swam ashore after his fall, and that the raiders overtook him on their retreat, and that he was now immured, a Federal prisoner. The still and all the effects of the brush-whiskey trade disappeared as mysteriously, and doubtless this silent flitting gave rise to the hopeful rumor that Tarbetts had been seen alive and well since that fateful night, and that in some farther recesses of the wilderness, undiscovered by the law, he and like comrades continue their chosen vocation. However that may be, the vicinity of Hoho-hebee Falls, always a lonely place, is now even a deeper solitude. The beavers, unmolested, haunt the ledges ; along their precipitous ways the deer come down to drink ; on bright days the rainbow hovers about the falls ; on bright nights they glimmer in the moon ; but never again have they glowed with the shoaling orange light of the furnace, intensifying to the deep tawny tints of its hot heart, like the rich glamours of some great topaz.

This alien glow it was thought had betrayed the place to the raiders, and Nehemiah's instrumentality was never discovered. The post-office appointment was bestowed upon his rival for the position, and it was thought somewhat strange that he should endure the defeat with such exemplary resignation. No one seemed to connect his candidacy with his bootless search for his nephew. When Leander chanced to be mentioned, however, he observed with some rancor that he reckoned it was just as well he didn't come up with Lee-yander; there was generally mighty little good in a runaway boy, and Lee-yander had the name of being disobejent an' turr'ble bad.

Leander found a warm welcome at home. His violin had been broken in the *mêlée*, and the miller, though ardently urged, never could remember the spot where he had hidden the book—such havoc had the confusion of that momentous night wrought in his mental processes. Therefore, unhampered by music or literature, Leander addressed himself to the plough-handles, and together that season he and " Neighbor " made the best crop of their lives.

Laurelia sighed for the violin and Leander's music, though, as she always made haste to say, some pious people misdoubted whether it were not a sinful pastime. On such occasions it went hard with Leander not to divulge his late experiences and the connection of the pious Uncle Nehemiah therewith. But he always remembered in time Laurelia's disability to receive confidences, being a woman, and consequently unable to keep a secret, and he desisted.

One day, however, when he and Ty Sudley, ploughing the corn, now knee-high, were pausing to rest in the turn-row, a few furrows apart, in an ebullition of filial feeling he told all that had befallen him in his absence. Ty Sudley, divided between wrath toward Nehemiah and quaking anxiety for the dangers that Leander had been constrained to run—*ex post facto* tremors, but none the less acute—felt moved now and then to complacence in his prodigy.

"So 'twar *you-uns* ez war smart enough ter slam the furnace door an' throw the whole place inter darkness! That saved them moonshiners and raiders from killin' each other. It saved a deal o' bloodshed—ez sure ez shootin'. 'Twar mighty smart in ye. But"—suddenly bethinking himself of sundry unfilial gibes at Uncle Nehemiah and the facetious account of his plight—"Lee-yander, ye mustn't be so turr'ble bad, sonny; ye *mustn't* be so *turr'ble* bad."

"Naw, ma'am, Neighbor, I won't," Leander protested.

And he went on following the plough down the furrow and singing loud and clear.

THE RIDDLE OF THE ROCKS

UPON the steep slope of a certain " bald " among the Great Smoky Mountains there lie, just at the verge of the strange stunted woods from which the treeless dome emerges to touch the clouds, two great tilted blocks of sandstone. They are of marked regularity of shape, as square as if hewn with a chisel. Both are splintered and fissured; one is broken in twain. No other rock is near. The earth in which they are embedded is the rich black soil not unfrequently found upon the summits. Nevertheless no great significance might seem to attach to their isolation — an outcropping of ledges, perhaps ; a fracture of the freeze ; a trace of ancient denudation by the waters of the spring in the gap, flowing now down the trough of the gorge in a silvery braid of currents, and with a murmur that is earnest of a song.

It may have been some distortion of the story heard only from the lips of the circuit rider, some fantasy of tradition invested with the urgency of fact, but Roger Purdee could not remember the time when he did not believe that these were the stone tables of the Law that Moses flung down from the mountain-top in his wrath. In the dense ignorance of the mountaineer, and his secluded life, he knew of no foreign countries, no land holier than the land of his home. There was no incongruity to his mind that it should have been in the solemn silence and austere solitude of the " bald," in the magnificent ascendency of the Great Smoky, that the law-giver had met the Lord and spoken with Him. Often as he lay at length on the strange barren place, veiled with the clouds that frequented it, a sudden sunburst in their midst would suggest anew what supernal splendors had once been here vouchsafed to the faltering eye of man. The illusion had come to be very dear to him; in this insistent localization of his faith it was all very near. And so he would go down to the slope below, among the weird, stunted trees, and look once more upon the broken tables, and ponder upon the strange signs written by time thereon. The insistent fall of the rain, the incisive blasts of the wind, coming again and again, though the centuries went, were registered here in mystic runes. The surface had weathered to a whitish-gray, but still in tiny depressions its pristine dark color showed in rugose characters. A splintered fissure held delicate fucoid impressions in fine script full of meaning. A series of worm-holes traced erratic hieroglyphics across a scaling corner; all the varied texts were illuminated by quartzose particles glittering in the sun, and here and there fine green grains of glauconite. He knew no names like these, and naught of meteorological potency. He had studied no other rock. His casual notice had been arrested nowhere by similar signs. Under the influence of his ignorant superstition, his cherished illusion, the lonely wilderness,

what wonder that, as he pondered upon the rocks,
strange mysteries seemed revealed to him? He
found significance in these cabalistic scriptures—
nay, he read inspired words! With the ramrod of
his gun he sought to follow the fine tracings of the
letters writ by the finger of the Lord on the stone
tables that Moses flung down from the mountain-
top in his wrath.

With a devout thankfulness Purdee realized that he
owned the land where they lay. It was worth, perhaps, a few cents an acre; it was utterly untillable,
almost inaccessible, and his gratulation owed its
fervor only to its spiritual values. He was an idle
and shiftless fellow, and had known no glow of acquisition, no other pride of possession. He herded
cattle much of the time in the summer, and he hunted
in the winter—wolves chiefly, their hair being long
and finer at this season, and the smaller furry gentry; for he dealt in peltry. And so, despite the
vastness of the mountain wilds, he often came and
knelt beside the rocks with his rifle in his hand,
and sought anew to decipher the mystic legends.
His face, bending over the tables of the Law with
the earnest research of a student, with the chastened
subduement of devotion, with all the calm sentiments
of reverie, lacked something of its normal aspect.
When a sudden stir of the leaves or the breaking of
a twig recalled him to the world, and he would lift
his head, it might hardly seem the same face, so
heavy was the lower jaw, so insistent and coercive
his eye. But if he took off his hat to place therein
his cotton bandana handkerchief or (if he were in
luck and burdened with game) the scalp of a wild-

cat—valuable for the bounty offered by the State—
he showed a broad, massive forehead that added
the complement of expression, and suggested a
doubt if it were ferocity his countenance bespoke or
force.   His long black hair hung to his shoulders,
and he wore a tangled black beard ; his deep-set
dark blue eyes were kindled with the fires of imagination.   He was tall, and of a commanding presence but for his stoop and his slouch.   His garments
seemed a trifle less well ordered than those of his
class, and bore here and there the traces of the
blood of beasts ; on his trousers were grass stains
deeply grounded, for he knelt often to get a shot,
and in meditation beside the rocks.   He spent little
time otherwise upon his knees, and perhaps it was
some intuition of this fact that roused the wrath of
certain brethren of the camp-meeting when he suddenly appeared among them, arrogating to himself
peculiar spiritual experiences, proclaiming that his
mind had been opened to strange lore, repeating
thrilling, quickening words that he declared he had
read on the dead rocks whereon were graven the
commandments of the Lord.   The tumultuous tide
of his rude eloquence, his wild imagery, his ecstasy
of faith, rolled over the assembly and awoke it
anew to enthusiasms.   Much that he said was accepted by the more intelligent ministers who led
the meeting as figurative, as the finer fervors of
truth, and they felt the responsive glow of emotion
and quiver of sympathy.   He intended it in its simple, literal significance.   And to the more local
members of the congregation the fact was patent.

    " Sech a pack o' lies hev seldom been tole in the

hearin' o' Almighty Gawd," said Job Grinnell, a few days after the breaking up of camp. He was rehearsing the proceedings at the meeting partly for the joy of hearing himself talk, and partly at the instance of his wife, who had been prevented from attending by the inopportune illness of one of the children. "Ez I loant my ear ter the words o' that thar brazen buzzard I eyed him constant. Fur I looked ter see the jedgmint o' the Lord descend upon him like S'phira an' An'ias."

" *Who?*" asked his wife, pausing in her task of picking up chips. He had spoken of them so familiarly that one might imagine they lived close by in the cove.

"An'ias an' S'phira—them in the Bible ez war streck by lightnin' fur lyin'," he explained.

"I 'member *her*," she said. " S'phia, I calls her."

"Waal, A'gusta, *S'phira* do me jes ez well," he said, with the momentary sulkiness of one corrected. "Thar war a man along, though. An' 'pears ter me thar war powerful leetle jestice in thar takin' off, ef Roger Purdee be 'lowed ter stan' up thar in the face o' the meetin' an' lie so ez no yearthly critter in the worl' could b'lieve him—'ceptin' Brother Jacob Page, ez 'peared plumb out'n his head with religion, an' got ter shoutin' when this Purdee tuk ter tellin' the law he read on them rocks—Moses' tables, folks calls 'em—up yander in the mounting."

He nodded upward toward the great looming range above them. His house was on a spur of the mountain, overshadowed by it; shielded. It was to him the Almoner of Fate. One by one it doled out the days, dawning from its summit; and thence, too, came the darkness and the glooms of night. One by one it liberated from the enmesh-ments of its tangled wooded heights the constella-tions to gladden the eye and lure the fancy. Its largess of silver torrents flung down its slopes made fertile the little fields, and bestowed a lilting song on the silence, and took a turn at the mill-wheel, and did not disdain the thirst of the humble cattle. It gave pasturage in summer, and shelter from the winds of the winter. It was the assertive feature of his life ; he could hardly have imagined existence without " the mounting."

"Tole what he read on them rocks—yes, sir, ez glib ez swallerin' a persimmon. 'Twarn't the reg'lar ten comman'ments—some cur'ous new texts—jes a-rollin' 'em out ez sanctified ez ef he hed been called ter preach the gospel ! An' thar war Brother Eden Bates a-answerin' 'Amen' ter every one. An' Brother Jacob Page : 'Glory, brother ! Ye hev re-ceived the outpourin' of the Sperit ! Shake hands, brother !' An' sech ez that. Ter hev hearn the commotion they raised about that thar derned lyin' sinner ye'd hev 'lowed the meetin' war held ter glorify him stiddier the Lord."

Job Grinnell himself was a most notorious Chris-tian. Renown, however, with him could never be a superfluity, or even a sufficiency, and he grudged the fame that these strange spiritual utterances were acquiring. He had long enjoyed the distinc-tion of being considered a miraculous convert; his rescue from the wily enticements of Satan had been celebrated with much shaking and clapping of hands, and cries of "Glory," and muscular ecstasy.

His religious experiences thenceforth, his vacillations of hope and despair, had been often elaborated amongst the brethren. But his was a conventional soul; its expression was in the formulæ and platitudes of the camp-meeting. They sank into oblivion in the excitement attendant upon Purdee's wild utterances from the mystic script of the rocks.

As Grinnell talked, he often paused in his work to imitate the gesticulatory enthusiasms of the saints at the camp-meeting. He was a thickset fellow of only medium height, and was called, somewhat invidiously, "a chunky man." His face was broad, prosaic, good-natured, incapable of any fine gradations of expression. It indicated an elementary rage or a sluggish placidity. He had a ragged beard of a reddish hue, and hair a shade lighter. He wore blue jeans trousers and an unbleached cotton shirt, and the whole system depended on one suspender. He was engaged in skimming a great kettle of boiling sorghum with a perforated gourd, which caught the scum and strained the liquor. The process was primitive, instead of the usual sorghum boiler and furnace, the kettle was propped upon stones laid together so as to concentrate the heat of the fire. His wife was continually feeding the flames with chips which she brought in her apron from the wood-pile. Her countenance was half hidden in her faded pink sun-bonnet, which, however, did not obscure an expression responsive to that on the man's face. She did not grudge Purdee the salvation he had found; she only grudged him the prestige he had derived from its unique method.

"Why can't the critter elude Satan with less n'ise?" she asked, acrimoniously.

"Edzackly," her husband chimed in.

Now and then both turned a supervisory glance at the sorghum mill down the slope at some little distance, and close to the river. It had been a long day for the old white mare, still trudging round and round the mill; perhaps a long day as well for the two half-grown boys, one of whom fed the machine, thrusting into it a stalk at a time, while the other brought in his arms fresh supplies from the great pile of sorghum cane hard by.

All the door-yard of the little log cabin was bedaubed with the scum of the sorghum which Job Grinnell flung from his perforated gourd upon the ground. The idle dogs—and there were many—would find, when at last disposed to move, a clog upon their nimble feet. They often sat down with a wrinkling of brows and a puzzled expression of muzzle to investigate their gelatinous paws with their tongues, not without certain indications of pleasure, for the sorghum was very sweet; some of them, that had acquired the taste for it from imitating the children, openly begged.

One, a gaunt hound, hardly seemed so idle; he had a purpose in life, if it might not be called a profession. He lay at length, his paws stretched out before him, his head upon them; his big brown eyes were closed only at intervals; ever and again they opened watchfully at the movement of a small child, ten months old, perhaps, dressed in pink calico, who sat in the shadow formed by the protruding clay and stick chimney, and played by bouncing up and down and waving her fat hands, which seemed a perpetual joy and delight of possession to her. Take her altogether, she was a person of prepossessing appearance, despite her frank display of toothless gums, and around her wide mouth the unseemly traces of sorghum. She had the plumpest graces of dimples in every direction, big blue eyes with long lashes, the whitest possible skin, and an extraordinary pair of pink feet, which she rubbed together in moments of joy as if she had mistaken them for her hands. Although she sputtered a good deal, she had a charming, unaffected laugh, with the giggle attachment natural to the young of her sex.

Suddenly there sounded an echo of it, as it were —a shrill, nervous little whinny; the boys whirled round to see whence it came. The persistent rasping noise of the sorghum mill and the bubbling of the caldron had prevented them from hearing an approach. There, quite close at hand, peering through the rails of the fence, was a little girl of seven or eight years of age.

"I wanter kem in an' see you-uns's baby!" she exclaimed, in a high, shrill voice. "I want to pat it on the head."

She was a forlorn little specimen, very thin and sharp-featured. Her homespun dress was short enough to show how fragile were the long lean legs that supported her. The curtain of her sun-bonnet, which was evidently made for a much larger person, hung down nearly to the hem of her skirt; as she turned and glanced anxiously down the road, evidently suspecting a pursuer, she looked like an erratic sun-bonnet out for a stroll on a pair of borrowed legs.

She turned again suddenly and applied her thin, freckled little face to the crack between the rails. She smiled upon the baby, who smiled in response, and gave a little bounce that might be accounted a courtesy. The younger of the boys left the cane pile and ran up to his brother at the mill, which was close to the fence. "Don't ye let her do it," he said, venomously. "That thar gal is one of the Purdee fambly. I know her. Don't let her in." And he ran back to the cane.

Grinnell had seemed pleased by this homage at the shrine of the family idol; but at the very mention of the "Purdee fambly" his face hardened, an angry light sprang into his eyes, and his gesture in skimming with the perforated gourd the scum from the boiling sorghum was as energetic as if with the action he were dashing the "Purdee fambly" from off the face of the earth. It was an ancient feud; his grandfather and some contemporary Purdee had fallen out about the ownership of certain vagrant cattle; there had been blows and bloodshed; other members of the connection had been dragged into the controversy; summary reprisals were followed by counter-reprisals. Barns were mysteriously fired, hen-roosts robbed, horses unaccountably lamed, sheep feloniously sheared by unknown parties; the feeling widened and deepened, and had been handed down to the present generation with now and then a fresh provocation, on the part of one or the other, to renew and continue the rankling old grudges.

And here stood the hereditary enemy, wanting to pat their baby on the head.

"Naw, sir, ye won't!" exclaimed the boy at the mill, greatly incensed at the boldness of this proposition, glaring at the lean, tender, wistful little face between the rails of the fence.

But the baby, who had not sense enough to know anything about hereditary enemies, bounced and laughed and gurgled and sputtered with glee, and waved her hands, and had never looked fatter or more beguiling.

"I jes wanter pat it wunst," sighed the hereditary enemy, with a lithe writhing of her thin little anatomy in the anguish of denial—"*jes wunst!*"

"Naw, sir!" exclaimed the youthful Grinnell, more insistently than before. He did not continue, for suddenly there came running-down the road a boy of his own size, out of breath, and red and angry— the pursuer, evidently, that the hereditary enemy had feared, for she crouched up against the fence with a whimper.

"Kem along away from thar, ye miser'ble little stack o' bones!" he cried, seizing his sister by one hand and giving her a jerk—"a-foolin' round them Grinnells' fence an' a-hankerin' arter thar old baby!"

He felt that the pride of the Purdee family was involved in this admission of envy.

"I jes wanter pat it on the head *wunst*," she sighed.

"Waal, ye won't now," said the Grinnell boys in chorus.

The Purdee grasp was gentler on the little girl's arm. This was due not to fraternal feeling so much as to loyalty to the clan ; " stack o' bones " though she was, they were Purdee bones.

" Kem along," Ab Purdee exhorted her.   " A baby ain't nuthin' extry, nohow "—he glanced scoffingly at the infantile Grinnell.   " The mountings air fairly a-roamin' with 'em."

" We-uns 'ain't got none at our house," whined the sun-bonnet, droopingly, moving off slowly on its legs, which, indeed, seemed borrowed, so unsteady and loath to go they were.

The Grinnell boys laughed aloud, jeeringly and ostentatiously, and the Purdee blood was moved to retort : "We-uns don't want none sech ez that. Nary tooth in her head !"

And indeed the widely stretched babbling lips displayed a vast vacuity of gum.

Job Grinnell, who had listened with an attentive ear to the talk of the children, had nevertheless continued his constant skimming of the scum. Now he rose from his bent posture, tossed the scum upon the ground, and with the perforated gourd in his hand turned and looked at his wife. Augusta had dropped her apron and chips, and stood with folded arms across her breast, her face wearing an expression of exasperated expectancy.

The Grinnell boys were humbled and abashed. The wicked scion of the Purdee house, joying to note how true his shaft had sped, was again fitting his bow.

" An' ez bald-headed ez the mounting."

The baby had a big precedent, but although no peculiar shame attaches to the bare pinnacle of the summit, she—despite the difference in size and age

—was expected to show up more fully furnished, and in keeping with the rule of humanity and the gentilities of life.

No teeth, no hair, no sign of any: the fact that she was so backward was a sore point with all the family.   Job Grinnell suddenly dropped the perforated gourd, and started down toward the fence. The acrimony of the old feud was as a trait bred in the bone.   Such hatred as was inherent in him was evoked by his religious jealousies, and the pious sense that he was following the traditions of his elders and upholding the family honor blended in gentlest satisfaction with his personal animosity toward Roger Purdee as he noticed the boy edging off from the fence to a safe distance.   He eyed him derisively for a moment.

"Kin ye kerry a message straight?"   The boy looked up with an expression of sullen acquiescence, but said nothing. "Ax yer dad—an' ye kin tell him the word kems from me—whether he hev read sech ez this on the lawgiver's stone tables yander in the mounting: 'An' ye shall claim sech ez be yourn, an' yer neighbor's belongings shall ye in no wise boastfully medjure fur yourn, nor look upon it fur covetiousness, nor yit git up a big name in the kentry fur ownin' sech ez be another's.'"

He laughed silently—a twinkling, wrinkling demonstration over all his broad face—a laugh that was younger than the man, and would have befitted a square-faced boy.

The youthful Purdee, expectant of a cuffing, stood his ground more doubtfully still under the insidious thrusts of this strange weapon, sarcasm.   He knew that they were intended to hurt; he was wounded primarily in the intention, but the exact lesion he could not locate. He could meet a threat with a bold face, and return a blow with the best. But he was mortified in this failure of understanding, and perplexity cowed him as contention could not. He hung his head with its sullen questioning eyes, and he found great solace in a jagged bit of cloth on the torn bosom of his shirt, which he could turn in his embarrassed fingers.

"Whar be yer dad?" Grinnell asked.

"Up yander in the mounting," replied the subdued Purdee.

"A-readin' of mighty s'prisin' matter writ on the rocks o' the yearth!" exclaimed Grinnell, with a laugh. "Waal, jes keep that sayin' o' mine in yer head, an' tell him when he kems home. An' look a-hyar, ef enny mo' o' his stray shoats kem about hyar, I'll snip thar ears an' gin 'em my mark."

The youth of the Purdee clan meditated on this for a moment. He could not remember that they had missed any shoats. Then the full meaning of the phrase dawned upon him—it was he and the wiry little sister thus demeaned with a porcine appellation, and whose ears were threatened. He looked up at the fence, the little low house, the barn close by, the sorghum mill, the drying leaves of tobacco on the scaffold, the saltatory baby; his eyes filled with helpless tears, that could not conceal the burning hatred he was born to bear them all. He was hot and cold by turns; he stood staring, silent and defiant, motionless, sullen. He heard the melodic measure of the river, with its crystalline, keen vibrations against the rocks ; the munching teeth of the old mare—allowed to come to a stand-still that the noise of the sorghum mill might not impinge upon the privileges of the quarrel ; and the high, ecstatic whinny of the little sister waiting on the opposite bank of the river, having crossed the foot-bridge. There the Grinnell baby had chanced to spy her, and had bounced and grinned and sputtered affably.   It was she who had made all the trouble yearning after the Grinnell baby.

He would not stay, however, to be ignominiously beaten, for Grinnell had turned away, and was looking about the ground as if in search of a thick stick.   He accounted himself no craven, thus numerically at a disadvantage, to turn shortly about, take his way down the rocky slope, cross the foot-bridge, jerk the little girl by one hand and lead her whimpering off, while the round-eyed Grinnell baby stared gravely after her with inconceivable emotions.   These presently resulted in rendering her cross ; she whined a little and rubbed her eyes, and, smarting from her own ill-treatment of them, gave a sharp yelp of dismay.   The old dog arose and went and sat close by her, eying her solemnly and wagging his tail, as if begging her to observe how content he was.   His dignity was somewhat impaired by sudden abrupt snaps at flies, which caused her to wink, stare, and be silent in astonishment.

"Waal, Job Grinnell," exclaimed Augusta, as her husband came back and took the perforated gourd from her hand — for she had been skimming the

20

sorghum in his absence — "ye air the longest-tongued man, ter be so short-legged, I ever see!"

He looked a trifle discomfited. He had deported himself with unwonted decision, conscious that Augusta was looking on, and in truth somewhat supported by the expectation of her approval.

"What ails ye ter say words ye can't abide by—ye 'low ye 'pear so graceful on the back track?" she asked.

He bent over the sorghum, silently skimming. His composure was somewhat ruffled, and in throwing away the scum his gesture was of negligent and discursive aim ; the boiling fluid bespattered the foot of one of the omnipresent dogs, whose shrieks rent the sky and whose activity on three legs amazed the earth. He ran yelping to Mrs. Grinnell, nearly overturning her in his turbulent demand for sympathy ; then scampered across to the boys, who readily enough stopped their work to examine the wounded member and condole with its wheezing proprietor.

"What ye mean, A'gusta?" Grinnell said at length. "Kase I 'lowed I'd cut thar ears? I ain't foolin'. Kem meddlin' about remarkin' on our chill'n agin, I'll show 'em."

Augusta looked at him in exasperation. "I ain't keerin' ef all the Purdees war deef," she remarked, inhumanly, "but what war them words ye sent fur a message ter Purdee?—'bout pridin' on what ain't theirn."

Grinnell in his turn looked at her—but dubiously, However much a man is under the domination of his wife, he is seldom wholly frank. It is in this wise that his individuality is preserved to him. "I war jes wantin' ter know ef them words war on the rocks," he said with a disingenuousness worthy of a higher culture.

She received this with distrust. "I kin tell ye now—they ain't," she said, discriminatingly; "Purdee's words don't sound like *them*."

"Waal, now, what's the differ?" he demanded, with an indignation natural enough to aspiring humanity detecting a slur upon one's literary style.

"Waal—" she paused as she knelt down to feed the fire, holding the fragrant chips in her hand ; the flame flickered out and lighted up her reflective eyes while she endeavored to express the distinction she felt: "Purdee's words don't sound ter me like the words of a man sech ez men be."

Grinnell wrinkled his brows, trying to follow her here.

"They sound ter me like the words spoke in a dream—the pernouncings of a vision." Mrs. Grinnell fancied that she too had a gift of Biblical phraseology. "They sound ter me like things I hearn whenst I war a-hungered arter righteousness an' seekin' religion, an' bided alone in the wilderness a-waitin' o' the Sperit."

"'Gusta !" suddenly exclaimed her husband, with the cadence of amazed conviction, "ye b'lieve the lie o' that critter, an' that he reads the words o' the Lord on the rock !"

She looked up a little startled. She had been unconscious of the circuitous approaches of credence, and shared his astonishment in the conclusion.

"Waal, sir !" he said, more hurt and cast down

than one would have deemed possible. " I'm willin'
ter hev it so. I'm jes nuthin' but a sinner an' a fool,
ripenin' fur damnation, an' he air a saint o' the
yearth !"

Now such sayings as this were frequent upon Job
Grinnell's tongue. He did not believe them ; their
utility was in their challenge to contradiction. Thus
they often promoted an increased cordiality of the
domestic relations and an accession of self-esteem.

Augusta, however, was tired ; the boiling sorghum
and the September sun were debilitating in their
effects. There was something in the scene with the
youthful Purdee that grated upon her half-developed
sensibilities. The baby was whimpering outright,
and the cow was lowing at the bars. She gave her
irritation the luxury of withholding the salve to Grinnell's wounded vanity. She said nothing. The
tribute to Purdee went for what it was worth, and he
was forced to swallow the humble-pie he had taken
into his mouth, albeit it stuck in his throat.

A shadow seemed to have fallen into the moral
atmosphere as the gentle dusk came early on. One
had a sense as if bereft, remembering that so short
a time ago at this hour the sun was still high, and
that the full-pulsed summer day throbbed to a climax of color and bloom and redundant life. Now,
the scent of harvests was on the air ; in the stubble
of the sorghum patch she saw a quail's brood more
than half-grown, now afoot, and again taking to
wing with a loud whirring sound. The perfume of
ripening muscadines came from the bank of the
river. The papaws hung globular among the leaves
of the bushes, and the persimmons were reddening.

The vermilion sun was low in the sky above the purpling mountains; the stream had changed from a crystalline brown to red, to gold, and now it was beginning to be purple and silver. And this reminded her that the full-moon was up, and she turned to look at it — so pearly and luminous above the jagged ridge-pole of the dark little house on the rise. The sky about it was blue, refining into an exquisitely delicate and ethereal neutrality near the horizon. The baby had fallen asleep, with its bald head on the old dog's shoulder.

After the supper was over, the sorghum fire still burned beneath the great kettle, for the syrup was not yet made, and sorghum-boiling is an industry that cannot be intermitted. The fire in the midst of the gentle shadow and sheen of the night had a certain profane, discordant effect. Pete's ill-defined figure slouching over it while he skimmed the syrup was grimly suggestive of the distillations of strange elixirs and unhallowed liquors, and his simple face, lighted by a sudden darting red flame, had unrecognizable significance and was of sinister intent. For Pete was detailed to attend to the boiling; the grinding was done, and the old white mare stood still in the midst of the sorghum stubble and the moonlight, as motionless and white as if she were carved in marble. Job Grinnell sat and smoked on the porch.

Presently he got up suddenly, knocked the ashes out of his pipe, and looked at it carefully before he stuck it into his pocket. He went, without a word, down the rocky slope, past the old drowsing mare, and across the foot-bridge. Two or three of the

dogs, watching him as he reappeared on the opposite
bank, affected a mistake in identity. They growled,
then barked outright, and at last ran down and
climbed the fence and bounded about it, baying the
vista where he had vanished, until the sleepy old
mare turned her head and gazed in mild surprise at
them.

Augusta sat alone on the step of the porch.

She had various regrets in her mind, incipient
even before he had quite gone, and now defining
themselves momently with added poignancy.   A
woman who, in her retirement at home, charges herself with the control of a man's conduct abroad, is
never likely to be devoid of speculation upon probable disasters to ensue upon any abatement of the
activities of her discretion.   She was sorry that she
had allowed so trifling a matter to mar the serenity
of the family ; her conscience upbraided her that
she had not besought him to avoid the blacksmith's
shop, where certain men of the neighborhood were
wont to congregate and drink deep into the night.
Above all, her mind went back to the enigmatical
message, and she wondered that she could have
been so forgetful as to fail to urge him to forbear
angering Purdee, for this would have a cumulative
effect upon all the rancors of the old quarrels, and
inaugurate perhaps a new series of reprisals.

"I ain't afeard o' no Purdee ez ever stepped,"
she said to herself, defining her position.   " But I'm
fur peace.   An' ef the Purdees will leave we-uns be,
I ain't a-goin' ter meddle along o' them."

She remembered an old barn-burning, in the days
when she and her husband were newly married, at

his father's house. She looked up at the barn hard by, on a line with the dwelling, with that tenderness which one feels for a thing, not because of its value, but for the sake of possession, for the kinship with the objects that belong to the home. A cat was sitting high in a crevice in the logs where the daubing had fallen out ; the moon glittered in its great yellow eyes. A frog was leaping along the open space about the rude step at Augusta's feet. A clump of mullein leaves, silvered by the light, spangled by the dew, hid him presently. What an elusive glistening gauze hung over the valley far below, where the sense of distance was limited by the sense of sight !—for it was here only that the night, though so brilliant, must attest the incomparable lucidity of daylight. She could not even distinguish, amidst those soft sheens of the moon and the dew, the Lombardy poplar that grew above the door of old Squire Grove's house down in the cove ; in the daytime it was visible like a tiny finger pointing upward. How drowsy was the sound of the katydid, now loudening, now falling, now fainting away ! And the tree-toad shrilled in the dog-wood tree. The frogs, too, by the river in iterative fugue sent forth a song as suggestive of the margins as the scent of the fern, and the mint, and the fragrant weeds.

A convulsive start ! She did not know that she slept until she was again awake. The moon had travelled many a mile along the highways of the skies. It hung over the purple mountains, over the farthest valley. The cicada had grown dumb. The stars were few and faint. The air was chill.

She started to her feet; her garments were heavy with dew. The fire beneath the sorghum kettle had died to a coal, flaring or fading as the faint fluctuations of the wind might will. Near it Pete slumbered where he too had sat down to rest. And Job—Job had never returned.

He had found it a lightsome enough scene at the blacksmith's shop, where it was understood that the neighboring politicians collogued at times, or brethren in the church discussed matters of discipline or more spiritual affairs. In which of these interests a certain corpulent jug was most active it would be difficult perhaps to accurately judge. The great barn-like doors were flung wide open, and there was a group of men half within the shelter and half without; the shoeing-stool, a broken plough, an empty keg, a log, and a rickety chair sufficed to seat the company. The moonlight falling into the door showed the great slouching, darkling figures, the anvil, the fire of the forge (a dim ashy coal), and the shadowy hood merging indistinguishably into the deep duskiness of the interior. In contrast, the scene glimpsed through the low window at the back of the shop had a certain vivid illuminated effect. A spider web, revealing its geometric perfection, hung half across one corner of the rude casement; the moonbeams without were individualized in fine filar delicacy, like the ravellings of a silver skein. The boughs of a tree which grew on a slope close below almost touched the lintel; the leaves seemed a translucent green; a bird slept on a twig, its head beneath its wing.

THE BLACKSMITH'S SHOP

Back of the cabin, which was situated on a limited terrace, the great altitudes of the mountain rose into the infinity of the night.

The drawling conversation was beset, as it were, by faint fleckings of sound, lightly drawn from a crazy old fiddle under the chin of a gaunt, yellow-haired young giant, one Ephraim Blinks, who lolled on a log, and who by these vague harmonies unconsciously gave to the talk of his comrades a certain theatrical effect.

Grinnell slouched up and sat down among them, responding with a nod to the unceremonious "Hy're, Job?" of the blacksmith, who seemed thus to do the abbreviated honors of the occasion. The others did not so formally notice his coming.

The subject of conversation was the same that had pervaded his own thoughts. He was irritated to observe how Purdee had usurped public attention, and yet he himself listened with keenest interest.

"Waal," said the ponderous blacksmith, "I kin onderstan' mighty well ez Moses would hev been mighty mad ter see them folks a-worshippin' o' a calf—senseless critters they be! 'Twarn't no use flingin' down them rocks, though, an' gittin' 'em bruk. Sandstone ain't like metal; ye can't heat it an' draw it down an' weld it agin."

His round black head shone in the moonlight, glistening because of his habit of plunging it, by way of making his toilet, into the barrel of water where he tempered his steel. He crossed his huge folded bare arms over his breast, and leaned back against the door on two legs of the rickety chair.

"Naw, sir," another chimed in. "He mought hev knowed he'd jes hev ter go ter quarryin' agin."

"They air always a-crackin' up them folks in the Bible ez sech powerful wise men," said another, whose untrained mind evidently held the germs of advanced thinking. "'Pears ter me ez some of 'em conducted tharselves ez foolish ez enny folks I know —this hyar very Moses one o' 'em. Throwin' down them rocks 'minds me o' old man Pinner's tantrums. Sher'ff kem ter his house 'bout a jedgmint debt, an' levied on his craps. An' arter he war gone old man tuk a axe an' gashed bodaciously inter the loom an' hacked it up. Ez ef that war goin' ter do enny good! His wife war the mos' outed woman I ever see. They 'ain't got nare nother loom nuther, an' hain't hearn no advices from the Lord."

The violinist paused in his playing. "They 'lowed Moses war a meek man too," he said. "He killed a man with a brick-badge an' buried him in the sand. Mighty meek ways"—with a satirical grimace.

The others, divining that this was urged in justification and precedent for devious modern ways that were not meek, did not pursue this branch of the subject.

"S'prised me some," remarked the advanced thinker, "ter hear ez them tables o' stone war up on the bald o' the mounting thar. I hed drawed the idee ez 'twar in some other kentry somewhar—I dunno—" He stopped blankly. He could not formulate his geographical ignorance. "An' I never knowed," he resumed, presently, "ez thar war enough gold in Tennessee ter make a gold calf; they fund gold hyar, but 'twar mighty leetle."

"Mebbe 'twar a mighty leetle calf," suggested the blacksmith.

"Mebbe so," assented the other.

"Mebbe 'twar a silver one," speculated a third; "plenty o' silver they' low thar air in the mountings."

The violinist spoke up suddenly. "Git one o' them Injuns over yander ter Quallatown right seasonable drunk, an' he'll tell ye a power o' places whar the old folks said thar war silver." He bowed his chin once more upon the instrument, and again the slow drawling conversation proceeded to soft music.

"Ef ye'll b'lieve me," said the advanced thinker, "I never war so conflusticated in my life ez I war when he stood up in meetin' an' told 'bout'n the tables o' the law bein' on the bald! I 'lowed 'twar somewhar 'mongst some sort'n people named 'Gyptians.'"

"Mebbe some o' them Injuns air named 'Gyptians," suggested Spears, the blacksmith.

"Naw, sir," spoke up the fiddler, who had been to Quallatown, and was the ethnographic authority of the meeting. "Tennessee Injuns be named Cher'kee, an' Chick'saw, an' Creeks."

There was a silence. The moonlight sifted through the dark little shanty of a shop; the fretting and foaming of a mountain stream arose from far down the steep slope, where there was a series of cascades, a fine water-power, utilized by a mill. The sudden raucous note of a night-hawk jarred upon the air, and a shadow on silent wings sped past. The road was dusty in front of the shop, and for a

space there was no shade. Into the full radiance
of the moonlight a rabbit bounded along, rising
erect with a most human look of affright in its great
shining eyes as it tremulously gazed at the motionless figures. It too was motionless for a moment.
The young musician made a lunge at it with his
bow ; it sprang away with a violent start—its elongated grotesque shadow bounding kangaroo-like beside it—into the soft gloom of the bushes. There
was no other traveller along the road, and the talk
was renewed without further interruption. "Waal,
sir, ef 'twarn't fur the testimony o' the words he reads
ez air graven on them rocks, I couldn't git my
cornsent ter b'lieve ez Moses ever war in Tennessee," said the advanced thinker. "I ain't onder-
takin' ter say what State he settled in, but I 'lowed
'twarn't hyar. It mus' hev been, though, 'count o'
the scripture on them broken tables."

"I never knowed a meetin' woke ter sech a
pint o' holiness. The saints jes rampaged around
till it fairly sounded like the cavortin's o' the ungodly," a retrospective voice chimed in.

"I raised thirty - two hyme chunes," said the
musician, who had a great gift in quiring, and was
the famed possessor of a robust tenor voice. "A
leetle mo' gloryin' aroun' an' I'd hev kem ter the
eend o' my row, an' hev hed ter begin over agin."
He spoke with acrimony, reviewing the jeopardy
in which his *répertoire* had been placed.

"Waal," said the blacksmith, passing his hand
over his black head, as sleek and shining as a
beaver's, "I'm a-goin' up ter the bald o' the mounting some day soon, ef so be I kin make out ter

shoe that mare o' mine "—for the blacksmith's mount was always barefoot—" I'm afeard ter trest her unshod on them slippery slopes; I want ter read some o' them sayin's on the stone tables myself. I likes ter git a tex' or the eend o' a hyme set a-goin' in my head—seems somehow ter teach itself ter the anvil, an' then it jes says it back an'. forth all day. Yestiddy I never seen its beat— ' Christ—war—born—in—Bethlehem.' The anvil jes rang with that ez ef the actial metal hed the gift o' prayer an' praise."

" Waal, sir," exclaimed Job Grinnell, who had been having frequent colloquies aside with the companionable jug, " ye mought jes ez well save yer shoes an' let yer mare go barefoot. Thar ain't nare sign o' a word writ on them rocks."

They all sat staring at him. Even the singing, long-drawn vibrations of the violin were still.

" By Hokey !" exclaimed the young musician, " I'll take Purdee's word ez soon ez yourn."

The whiskey which Grinnell had drunk had rendered him more plastic still to jealousy. The day was not so long past when Purdee's oath would have been esteemed a poor dependence against the word of so zealous a brother as he—a pillar in the church, a shining light of the congregation. He noted the significant fact that it behooved him to justify himself; it irked him that this was exacted as a tribute to Purdee's newly acquired sanctity.

" Purdee's jes a-lyin' an' a-foolin' ye," he declared. " Ever been up on the bald ?"

They had lived in its shadow all their lives.

Even by the circuitous mountain ways it was not more than five miles from where they sat. But none had chanced to have a call to go, and it was to them as a foreign land to be explored.

"Waal, I hev, time an' agin," said Grinnell. "I dunno who gin them rocks the name of Moses' .tables o' the Law. Moses must hev.hed a powerful block an' tackle ter lift sech tremenjious rocks. I hev known 'em named sech fur many a year. But I seen 'em not three weeks ago, an' thar ain't nare word writ on 'em. Thar's the mounting; thar's the rocks; ye kin go an' stare-gaze 'em an' sati'fy yerse'fs."

Whether it were by reason of the cumulative influences of the continual references to the jug, or of that sense of reviviscence, that more alert energy, which the cool Southern nights always impart after the sultry summer days, the suggestion that they should go now and solve the mystery, and meet the dawn upon the summit of the bald, found instant acceptance, which it might not have secured in the stolid daylight.

The moon, splendid, a lustrous white encircled by a great halo of translucent green, swung high above the duskily purple mountains. Below in the valleys its progress was followed by an opalescent gossamer presence that was like the overflowing fulness, the surplusage, of light rather than mist. The shadows of the great trees were interlaced with dazzling silver gleams. The night was almost as bright as the day, but cool and dank, full of sylvan fragrance and restful silence and a romantic liberty.

The blacksmith carried his rifle, for wolves were

often abroad in the wilderness. Two or three others were similarly armed; the advanced thinker had a hunting-knife, Job Grinnell a pistol that went by the name of "shootin'-iron." The musician carried no weapon. "I ain't 'feared o' no wolf," he said; "I'll play 'em a chune." He went on in the vanguard, his tousled yellow hair idealized with many a shimmer in the moonlight as it hung curling down on his blue jeans coat, his cheek laid softly on the violin, the bow glancing back and forth as if strung with moonbeams as he played. The men woke the solemn silences with their loud mirthful voices; they startled precipitate echoes; they fell into disputes and wrangled loudly, and would have turned back if sure of the way home, but Job Grinnell led steadily on, and they were fain to follow. They lagged to look at a spot where some man, unheeded even by tradition, had dug his heart's grave in a vain search for precious metal. A deep excavation in the midst of the wilderness told the story; how long ago it was might be guessed from the age of a stalwart oak that had sunk roots into its depths; the shadows were heavy about it; a sense of despair brooded in the loneliness. And so up and up the endless ascent; sometimes great chasms were at one side, stretching further and further, and crowding the narrow path—the herder's trail—against the sheer ascent, till it seemed that the treacherous mountains were yawning to engulf them. The air was growing colder, but was exquisitely clear and exhilarating; the great dewy ferns flung silvery fronds athwart the way; vines in stupendous lengths swung from the tops of gigantic

trees to the roots.   Hark ! among them birds chirp ;
a matutinal impulse seems astir in the woods ; the
moon is undimmed ; the stars faint only because
of her splendors ; but one can feel that the earth
has roused itself to a sense of a new day.   And
there, with such feathery flashes of white foam, such
brilliant straight lengths of translucent water, such
a leaping grace of impetuous motion, the currents
of the mountain stream, like the arrows of Diana,
shoot down the slopes.   And now a vague mist is
among the trees, and when it clears away they seem
shrunken, as under a spell, to half their size.   They
grow smaller and smaller -still, oak and chestnut
and beech, but dwarfed and gnarled like some old
orchard.   And suddenly they cease, and the vast
grassy dome uprises against the sky, in which the
moon is paling into a dull similitude of itself ; no
longer wondrous, transcendent, but like some lily
of opaque whiteness, fair and fading.   Beneath is
a purple, deeply serious, and sombre earth, to which
mists minister, silent and solemn ; myriads of
mountains loom on every hand ; the half-seen mysteries of the river, which, charged with the red clay
of its banks, is of a tawny color, gleams as it winds
in and out among the white vapors that reach in
fantastic forms from heaven above to the valley
below.   There is a certain relief in the mist—it
veils the infinities of the scene, on which the mind
can lay but a trembling hold.

   "Folks tell all sort'n cur'ous tales 'bout'n this
hyar spot," said Job Grinnell, his square face, his
red hair hanging about his ears, and his ragged red
beard visible in the dull light of the coming day.

"I hev hearn folks 'low ez a pa'tridge up hyar will look ez big ez a Dominicky rooster. An' ef ye listens ye kin hear words from somewhar. An' sometimes in the cattle-herdin' season the beastises will kem an' crowd tergether, an' stan' on the bald in the moonlight all night."

"I dunno," said the advanced thinker, "ez I be s'prised enny ef Purdee, ez be huntin' up hyar so constant, hev got sorter teched in the head, ter take up sech a cur'ous notion 'bout'n them rocks."

He glanced along the slope at the spot, visible now, where Moses flung the stone tables and they broke in twain. And there, standing beside them, was a man of great height, dressed in blue jeans, his broad-brimmed hat pushed from his brow, and his meditative dark eyes fixed upon the rocks; a deer, all gray and antlered, lay dead at his feet, and his rifle rested on the ground as he leaned on the muzzle.

A glance was interchanged between the others. Their intention, the promptings of curiosity, had flagged during the long tramp and the gradual waning of the influence of the jug. The coincidence of meeting Purdee here revived their interest. Grinnell, remembering the ancient feud, held back, being unlikely to elicit Purdee's views in the face of their contradiction. The blacksmith and the young fiddler took their way down toward him.

He looked up with a start, seeing them at some little distance. His full, contemplative eyes rested upon them for a moment almost devoid of questioning. It was not the face of a man who finds himself confronted with the discovery of his duplicity and his hypocrisy. There was a strange doubt stirring in the blacksmith's heart. As he approached he looked upon the storied rocks with a sort of solemn awe, as if they had indeed been given by the hand of the Lord to his servant, who broke them here in his wrath. He knew that the step of the musician slackened as he followed. What holy mysteries were they not rushing in upon? He spoke in a bated voice.

"Roger," he said, "we'uns hearn ye tell 'bout the scriptures graven on these hyar tables ez Moses flung down, an' we'uns 'lowed we'uns would kem an' read some fur ourselves."

Purdee did not speak nor hesitate; he moved aside that the blacksmith might stand where he had been—as it were at the foot of the page.

. But what transcendent glories thronged the heavens — what august splendors of dawn! Had the sun ever before risen like this, with the sky an emblazonment of red, of gold, of darting gleams of light; with the mountains most royally purple or most radiantly blue; with the prismatic mists in flight; with the slow climax of the dazzling sphere ascending to dominate it all?

The blacksmith knelt down to read. The musician, his silent violin under his chin, leaned over his comrade's shoulder. The hunter stood still, expectant.

Alas! the corrugations of time; the fissile results of the frost; the wavering line of ripple-marks of seas that shall ebb no more; growth of lichen; an army of ants in full march; a passion-flower trailing from a crevice, its purple blooms lying upon

THE ANGEL OF THE LAW

the gray stone near where it is stamped with the fossil imprint of a sea-weed, faded long ago and forgotten. Or is it, alas! for the eyes that can see only this?

The blacksmith looked up with a twinkling leer; the violinist recovered his full height, and drew the bow dashingly across the strings; then let his arm fall.

"Roger," the blacksmith said, "dad-burned ef I kin read ennything hyar."

The young musician looked over his brawny shoulder in silence.

"Whar d'ye make out enny letters, Roger?" persisted Spears.

Purdee leaned over and eagerly pointed with his ramrod to a curious corrugation of the surface of the rock. Again the blacksmith bent down; the musician craned forward, his yellow hair hanging about his bronzed face.

"I hev been toler'ble well acquainted with the alphabit," said Spears, "fur goin' on thirty year an' better, an' I'll swar ter Heaven thar ain't nare sign of a letter thar."

Purdee stared at him in wild-eyed amazement for a moment. Then he flung himself upon his knees beside the great rock, and guiding his ramrod over the surface, he exclaimed, "Hyar, Spears; right hyar!"

The blacksmith was all incredulous as he lent himself to a new posture, and leaned forward to look with the languid indulgence of one who will not again entertain doubt.

"Nare A, nor B, nor C, nor none o' the fambly,"

he declared. "These hyar rocks ain't no Moses' tables sure enough; Moses never war in Tennessee. They be jes like enny other rock; an' thar ain't a word o' writin' on 'em."

He looked up with a curious questioning at Purdee's face—a strange face for a man detected in a falsehood, a trick. The deep-set eyes were wide as if straining for perception denied them. Despite the chill, rare air, great drops had started on his brow, and were falling upon his beard, and upon his hands. These strong hands were quivering; they hovered above the signs on the rocks. The mystic letters, the inspired words, where were they? Grope as he might, he could not find them. Alas! doubt and denial had climbed the mountain—the awful limitations of the more finite human creature—and his inspiration and the finer enthusiasms of the truth were dead.

Dead with a throe that was almost like a literal death. This—on this he had lived; the ether of ecstasy was the breath of his life. He clutched at the stained red handkerchief knotted about his throat as if he were suffocating; he tore it open as he swayed backward on his knees. He did not hear—or he did not heed—the laugh among the little crowd on the bald—satirical, rallying, zestful. He was deaf to the strains of the violin, jeeringly and jerkingly playing a foolish tune. It was growing fainter, for they had all turned about to betake themselves once more to the world below. He could have seen, had he cared to see, their bearded grinning faces peering through the stunted trees, as descending they came near the spot where he

had lavished the spiritual graces of his feeling, his
enthusiasm, his devotion, his earnest reaching for
something higher, for something holy, which had
refreshed his famished soul; had given to its dumb-
ness words; had erased the values of the years, of
the nations; had made him friends with Moses on
the "bald"; had revealed to him the finger of the
Lord on the stone.

He took no heed of his gestures, of which, indeed, he was unconscious. They were fine dramatically, and of great power, as he alternately rose to
his full height, beating his breast in despair, and
again sank upon his knees, with a pondering brow
and a searching eye, and a hovering, trembling
hand, striving to find the clew he had lost. They
might have impressed a more appreciative audi-
ence, but not one more entertained than the cluster
of men who looked and paused and leered in
amusement at one another, and thrust out satirical
tongues. Long after they had disappeared, the
strains of the violin could be heard, filling the
solemn, stricken, strangely stunted woods with a
grotesquely merry presence, hilarious and jeering.

Purdee found it possible to survive the destruction of illusions. Most of us do. It wrought in
him, however, the saturnine changes natural upon
the relinquishment of a dear and dead fantasy.
This ethereal entity is a more essential component
of happiness than one might imagine from the extreme tenuity of the conditions of its existence.
Purdee's fantasy may have been a poor thing, but,
although he could calmly enough close its eyes,

and straighten its limbs, and bury it decently from out the offended view of fact, he felt that he should mourn it in his heart as long as he should live. And he was bereaved.

There is a certain stage in every sorrow when it rejects sympathy. Purdee, always taciturn, grave, uncommunicative, was invested with an austere aloofness, and was hardly to be approached as he sat, silent and absent, brooding over the fire at his own home. When roused by some circumstance of the domestic routine, and it became apparent that his mood was not sullenness or anger, but simple and complete introversion, it added a dignity and suggested a remoteness that were yet less reassuring. His son, who stood in awe of him— not because of paternal severity, but because no boy could refrain from a worshipping respect for so miraculous a shot, a woodsman so subtly equipped with all elusive sylvan instincts and knowledge — forbore to break upon his meditations by the delivery of Grinnell's message. Nevertheless the consciousness of withholding it weighed heavily upon him. He only pretermitted it for a time, until a more receptive state of mind should warrant it. Day by day, however, he looked with eagerness when he came into the cabin in the evening to ascertain if his father were still seated in the chimney-corner silently smoking his pipe. Purdee had seldom remained at home so long at a time, and the boy had a daily fear that the gun on the primitive rack of deer antlers would be missing, and word left in the family that he had taken the trail up the mountain, and would return "'cord-

in' ter luck with the varmints." And thus Job
Grinnell's enigmatical message, that had the ring
of defiance, might remain indefinitely postponed.

Abner had not realized how long a time it had
been delayed, until one evening at the wood-pile, in
tossing off a great stick to hew into lengths for the
chimney-place, he noticed that thin ice had formed
in the moss and the dank cool shadows of the inter-
stices. "I tell ye now, winter air a-comin'," he observed. He stood leaning on his axe-handle and
looking down upon the scene so far below; for Purdee's house was perched half-way up on the mountain - side, and he could see over the world how it
fared as the sun went down. Far away upon the
levels of the valley of East Tennessee a golden haze
glittered resplendent, lying close upon an irradiated earth, and ever brightening toward the horizon, and it seemed as if the sun in sinking might
hope to fall in fairer spheres than the skies he had
left, for they were of a dun-color and an opaque
consistency. Only one horizontal rift gave glimpses
of a dazzling ochreous tint of indescribable brilliancy, from the focus of which the divergent light
was shed upon the western limits of the land.
Chilhowee, near at hand, was dark enough—a purplish garnet hue; but the scarlet of the sour-wood
gleamed in the cove; the hickory still flared gallantly yellow; the receding ranges to the north
and south were blue and more faintly azure. The
little log cabin stood with small fields about it, for
Purdee barely subsisted on the fruits of the soil,
and did not seek to profit. It had only one room,
with a loft above; the barn was a makeshift of

poles, badly chinked, and showing through the
crevices what scanty store there was of corn and
pumpkins. A black-and-white work-ox, that had
evidently no deficiency of ribs, stood outside of the
fence and gazed, a forlorn Tantalus, at these unat-
tainable dainties ; now and then a muttered low
escaped his lips. Nobody noticed him or sympa-
thized with him, except perhaps the little girl, who
had come out in her sun-bonnet to help her brother
bring in the fuel. He gruffly accepted her company, a little ashamed of her because she was a girl ;
since, however, there was no other boy by to laugh,
he permitted her the delusion that she was of as-
sistance.

As he paused to rest he reiterated, " Winter air
a-comin', I tell ye."

"D'ye reckon, Ab," she asked, in her high, thin
little voice, her hands full of chips and the basket
at her feet, "ez Grinnell's baby knows Chris'mus
air a-comin' ?"

He glowered at her as he leaned on the axe. " I
reckon Grinnell's old baby dunno B from Bull-foot,"
he declared, gruffly.

The recollection of the message came over him.
He had a pang of regret, remembering all the old
grudges against the Grinnells. They were re-
enforced by this irrepressible yearning after their
baby, this admission that they had aught which
was not essentially despicable. Nevertheless, he
suddenly saw a reason for the Grinnell baby's existence ; he loaded up both arms with the sticks of
wood, and, followed by the peripatetic sun-bonnet,
conscientiously weighed down with one billet, he

strode into the house, and let his burden fall with a mighty clatter in the corner of the chimney. The sun-bonnet staggered up and threw her stick on the top of the pile of wood.

Purdee, sitting silently smoking, glanced up at the noise. Abner took advantage of the momentary notice to claim, too, the attention of his mother. "I wish ye'd make Eunice quit talkin' 'bout the Grinnells' old baby, like she war actially demented —uglies' bald-headed, slab-sided, slobbery old baby I ever see—nare tooth in its head! I do despise them Grinnells."

As he anticipated, his father spoke suddenly: "Ye jes keep away from thar," he said, sternly. "I trest them folks no furder 'n a rattlesnake."

"*I* ain't consortin' along o' 'em," declared the boy. "But I actially hed ter take Eunice by the scalp o' her head an' lug her off one day when she hung on thar fence a-stare-gazin' Grinnell's baby like 'twar fitten ter eat."

The child's mother, a cadaverous, pale woman, was listlessly stringing the warping-bars with hanks of variegated yarn. The grandmother, who conserved a much more active and youthful interest in life, took down a brown gourd used as a scrap-basket that was on a protruding lath of the clay-and-stick chimney, and hunted among the scraps of homespun and bits of yarn stowed within it. The room was much like the gourd in its aged brown tint; its indigenous aspect, as if it had not been made with hands, but was some spontaneous production of the soil; with its bits of bright color— the peppers hanging from the rafters, the rainbow-hued yarn festooning the warping-bars, the red coals of the fire, the blue and yellow ware ranged on the shelf, the brown puncheon floor and walls and ceiling and chimney—it might have seemed the interior of a similar gourd of gigantic proportions. She dressed a twig from the pile of wood in a gay scrap of cloth, casting glances the while at the little girl, and handed it to her.

"I hain't never seen ez good a baby ez this," she said, with the convincing coercive mendacity of a grandmother.

The little girl accepted it humbly; it was a good baby doubtless of its sort, but it was not alive, which could not be denied of the Grinnell baby, Grinnell though it was.

"An' Job Grinnell he kem down ter the fence, an' 'lowed he'd slit our ears, an' named us shoats," continued her brother. Purdee lifted his head. "An' sent a word ter dad," said the boy, tremulously.

"What word did he send ter—*me?*" cried Purdee.

The boy quailed to tell him. "He tole me ter ax ye ef ye ever read sech ez this on Moses' tables in the mountings—' An' ye shell claim sech ez be yer own, an' yer neighbors' belongings shell ye in no wise boastfully medjure fur yourn, nor look upon it fur covetiousness, nor yit git a big name up in the kentry fur ownin' sech ez be another's,' " faltered the sturdy Abner.

The next moment he felt an infinite relief. He suddenly recognized the fact that he had been chiefly restrained from repeating the words by an

WHAT WORD DID HE SEND TER— V? ...

unrealized terror lest they prove true—lest something his father claimed was not his, indeed.

But the expression of anger on Purdee's face was merged first in blank astonishment, then in perplexed cogitation, then in renewed and overpowering amazement.

The wife turned from the warping-bars with a vague stare of surprise, one hand poised uncertainly upon a peg of the frame, the other holding a hank of "spun truck." The grandmother looked over her spectacles with eyes sharp enough to seem subsidized to see through the mystery.

"In the name o' reason and religion, Roger Purdee," she adjured him, "what air that thar perverted Philistine talkin' 'bout?"

"It air more'n I kin jedge of," said Purdee, still vainly cogitating.

He sat for a time silent, his dark eyes bent on the fire, his broad, high forehead covered by his hat pulled down over it, his long, tangled, dark locks hanging on his collar.

Suddenly he rose, took down his gun, and started toward the door.

"Roger," cried his wife, shrilly, "I'd leave the critter be. Lord knows thar's been enough blood spilt an' good shelter burned along o' them Purdees' an' Grinnells' quar'ls in times gone. Laws-a-massy!"—she wrung her hands, all hampered though they were in the "spun truck"—"I'd ruther be a sheep 'thout a soul, an' live in peace."

"A sca'ce ch'ice," commented her mother. "Sheep's got ter be butchered. I'd ruther be the butcher, myself—healthier."

Purdee was gone.   He had glanced absently at
his wife as if he hardly heard.   He waited till she
paused ; then, without answer, he stepped hastily
out of the door and walked away.

The cronies at the blacksmith's shop latterly
gathered within the great flaring door, for the
frost lay on the dead leaves without, the stars
scintillated with chill suggestions, and the wind
was abroad on nights like these.   On shrill pipes
it played ; so weird, so wild, so prophetic were its
tones that it found only a shrinking in the heart of
him whose ear it constrained to listen.   The sound
of the torrent far below was accelerated to an
agitated, tumultuous plaint, all unknown when its
pulses were bated by summer languors.   The moon
was in the turmoil of the clouds, which, routed in
some wild combat with the winds, were streaming
westward.

And although the rigors of the winter were in
abeyance, and the late purple aster called the
Christmas - flower bloomed in the sheltered grass
at the door, the forge fire, flaring or dully glowing,
overhung with its dusky hood, was a friendly thing
to see, and in its vague illumination the rude interior of the shanty—the walls, the implements of
the trade, the bearded faces grouped about, the
shadowy figures seated on whatever might serve,
a block of wood, the shoeing - stool, a plough, or
perched on the anvil — became visible to Roger
Purdee from far down the road as he approached.
Even the head of a horse could be seen thrust in
at the window, while the brute, hitched outside, be-

guiled the dreary waiting by watching with a luminous, intelligent eye the gossips within, as if he understood the drawling colloquy. They were suffering some dearth of timely topics, supplying the deficiency with reminiscences more or less stale, and had expected no such sensation as they experienced when a long shadow fell athwart the doorway, —the broad aperture glimmering a silvery gray contrasted with the brown duskiness of the interior and the purple darkness of the distance ; the forge fire showed Purdee's tall figure leaning on the doorframe, and lighted up his serious face beneath his great broad-brimmed hat, his intent, earnest eyes, his tangled black beard and locks. He gave no greeting, and silence fell upon them as his searching gaze scanned them one by one.

"Whar's Job Grinnell?" he demanded, abruptly.

There was a shuffling of feet, as if those members most experienced relief from the constraint that silence had imposed upon the party. A vibration from the violin — a sigh as if the instrument had been suddenly moved rather than a touch upon the strings—intimated that the young musician was astir. But it was Spears, the blacksmith, who spoke.

"Kem in, Roger," he called out, cordially, as he rose, his massive figure and his sleek head showing in the dull red light on the other side of the anvil, his bare arms folded across his chest. "Naw, Job ain't hyar; hain't been hyar for a right smart while."

There was a suggestion of disappointment in the attitude of the motionless figure at the door. The

deeply earnest, pondering face, visible albeit the
red light from the forge-fire was so dull, was keenly
watched.   For the inquiry was fraught with peculiar meaning to those cognizant of the long and
bitter feud.

"I ax," said Purdee, presently, "kase Grinnell
sent me a mighty cur'ous word the t'other day."
He lifted his head.   "Hev enny o' you-uns hearn
him 'low lately ez I claim ennything ez ain't
mine ?"

There was silence for a moment.   Then the
forge was suddenly throbbing with the zigzagging
of the bow of the violin jauntily dandering along
the strings.   His keen sensibility apprehended the
sudden jocosity as a jeer, but before he could say
aught the blacksmith had undertaken to reply.

"Waal, Purdee, ef ye hedn't axed me, I warn't
layin' off ter say nuthin 'bout'n it.   'Tain't no consarn o' mine ez I knows on.   But sence ye *hev*
axed me, I hold my jaw fur the fear o' no man.
The words ain't writ ez I be feared ter pernounce.
An' ez all the kentry hev hearn 'bout'n it 'ceptin'
you-uns, I dunno ez I hev enny call ter hold my
jaw.   The Lord 'ain't set no seal on my lips ez I
knows on."

"Naw, sir !" said Purdee, his great eyes glooming through the dusk and flashing with impatience.
"He 'ain't set no seal on yer lips, ter jedge by the
way ye wallop yer tongue about inside o' 'em with
fool words.   Whyn't ye bite off what ye air tryin'
ter chaw ?"

"Waal, then," said the admonished orator, blunt-
ly, "Grinnell 'lows ye don't own that thar lan'

around them rocks on the bald, no more'n ye read
enny writin' on 'em."

"Not them rocks!" cried Purdee, standing suddenly erect—"the tables o' the Law, writ with the
finger o' the Lord—an' Moses flung 'em down thar
an' bruk 'em. All the kentry knows they air Moses'
tables. An' the groun' whar they lie air mine."

"'Tain't, Grinnell say 'tain't."

"Naw, sir," chimed in the young musician, his
violin silent. "Job Grinnell declars he owns it
hisself, an' ef he war willin' ter stan' the expense
he'd set up his rights, but the lan' ain't wuth it.
He 'lows his line runs spang over them rocks, an' a
heap furder."

Purdee was silent; one or two of the gossips
laughed jeeringly; he had been proved a liar once.
It was well that he did not deny; he was put to
open shame among them.

"An' Grinnell say," continued Blinks, "ez ye hev
gone an' tole big tales 'mongst the brethren fur
ownin' sech ez ain't yourn, an' readin' of s'prisin'
sayin's on the rocks."

He bent his head to a series of laughing har-
monics, and when he raised it, hearing no retort,
the silvery gray square of the door was empty.
He saw the moon glimmer on the clumps of grass
outside where the Christmas-flower bloomed.

The group sat staring in amaze; the blacksmith
strode to the door and looked out, himself a massive,
dark silhouette upon the shimmering neutrality of
the background. There was no figure in sight; no
faint foot-fall was audible, no rustle of the sere
leaves; only the voice of the mountain torrent, far

below, challenged the stillness with its insistent cry.

He looked back for a moment, with a vague, strange doubt if he had seen aught, heard aught, in the scene just past. " Hain't Purdee been hyar?" he asked, passing his hand across his eyes. The sense of having dreamed was so strong upon him that he stretched his arms and yawned.

The gleaming teeth of the grouped shadows demonstrated the merriment evoked by the query. The chuckle was arrested midway.

"Ye 'pear ter 'low ez suthin' hev happened ter Purdee, an' that thar war his harnt," suggested one.

The bold young musician laid down his violin suddenly. The instrument struck upon a keg of nails, and gave out an abrupt, discordant jangle, startling to the nerves. "Shet up, ye durned squeech-owl!" he exclaimed, irritably. Then, lowering his voice, he asked : " Didn't they 'low down yander in the Cove ez Widder Peters, the day her husband war killed by the landslide up in the mounting, heard a hoe a-scrapin' mightily on the gravel in the gyarden-spot, an' went ter the door, an' seen him thar a-workin', an' axed him when he kem home? An' he never lifted his head, but hoed on. An' she went down thar 'mongst the corn, an' she couldn't find nobody. An' jes then the Johns boys rid up an' 'lowed ez Jim Peters war dead, an' hed been fund in the mounting, an' they war a-fetchin' of him then."

The horse's head within the window nodded vi-

olently among the shadows, and the stones rolled
beneath his hoof as he pawed the ground.

"Mis' Peters she knowed suthin' were a-goin' ter
happen when she seen that harnt a-hoein'."

"I reckon she did," said the blacksmith, stretching himself, his nerves still under the delusion of
recent awakening. "Jim never hoed none when
he war alive. She mought hev knowed he war
dead ef she seen him hoein'."

"Waal, sir," exclaimed the violinist, "I'm a-goin'
up yander ter Purdee's ter-morrer ter find out what
he died of, an' when."

That he was alive was proved the next day, to
the astonishment of the smith and his friends.
The forge was the voting-place of the district, and
there, while the fire was flaring, the bellows blowing, the anvil ringing, the echo vibrating, now loud,
now faint, with the antiphonal chant of the hammer and the sledge, a notice was posted to inform
the adjacent owners that Roger Purdee's land, held
under an original grant from the State, would be
processioned according to law some twenty days
after date, and the boundaries thereof defined and
established. The fac-simile of the notice, too, was
posted on the court-house door in the county town
twenty miles away, for there were those who journeyed so far to see it.

"I wonder," said the blacksmith, as he stood in
the unfamiliar street and gazed at it, his big arms,
usually bare, now hampered with his coat sleeves
and folded upon his chest—"I wonder ef he footed
it all the way ter town at the gait he tuk when he
lit out from the forge?"

22

It was a momentous day when the county surveyor planted his Jacob's-staff upon the State line
on the summit of the bald.   His sworn chain-bearers, two tall young fellows clad in jeans, with
broad-brimmed wool hats, their heavy boots drawn
high over their trousers, stood ready and waiting,
with the sticks and clanking chain, on the margin
of the ice-cold spring gushing out on this bleak
height, and signifying more than a fountain in the
wilderness, since it served to define the southeast
corner of Purdee's land.   The two enemies were
perceptibly conscious of each other.   Grinnell's
broad face and small eyes-laden with fat lids were
persistently averted.   Purdee often glanced toward him gloweringly, his head held, nevertheless,
a little askance, as if he rejected the very sight.
There was the fire of a desperate intention in his
eyes.   Looking at his face, shaded by his broad-
brimmed hat, one could hardly have doubted now
whether it expressed most ferocity or force.   His
breath came quick — the bated breath of a man
who watches and waits for a supreme moment.
His blue jeans coat was buttoned close about his
sun-burned throat, where the stained red handkerchief was knotted.   He wore a belt with his pow-
·der-horn and bullet-pouch, and carried his rifle on
his shoulder; the hand that held it trembled, and
he tried to quell the quiver.   "I'll prove it fust,
an' kill him arterward — kill him arterward," he
muttered.

    In the other hand he held a yellowed old paper.
Now and then he bent his earnest dark eyes upon
the grant, made many a year ago by the State of

Tennessee to his grandfather; for there had been no subsequent conveyances.

The blacksmith had come begirt with his leather apron, his shirt-sleeves rolled up, and with his hammer in his hand, an inopportune customer having jeopardized his chance of sharing in the sensation of the day. The other neighbors all wore their coats closely buttoned. Blinks carried his violin hung upon his back; the sharp timbre of the wind, cutting through the leafless boughs of the stunted woods, had a kindred fibrous resonance. Clouds hung low far beneath them; here and there, as they looked, the trees on the slopes showed above and again below the masses of clinging vapors. Sometimes close at hand a peak would reveal itself, asserting the solemn vicinage of the place, then draw its veil slowly about it, and stand invisible and in austere silence. The surveyor, a stalwart figure, his closely buttoned coat giving him a military aspect, looked disconsolately downward.

"I hoped I'd die before this," he remarked. "I'm equal to getting over anything in nature that's flat or oblique, but the vertical beats me."

He bent to take sight for a moment, the group silently watching him. Suddenly he came to the perpendicular, and strode off down the rugged slope over gullies and bowlders, through rills and briery tangles, his eyes distended and eager as if he were led into the sylvan depths by the lure of a vision. The chain-bearers followed, continually bending and rising, the recurrent genuflections resembling the fervors of some religious rite. The chain rustled sibilantly among the dead leaves, and was ever and anon drawn out to its extremest length. Then the dull clank of the links was silent.

"Stick!" called out the young mountaineer in the rear.

"Stuck!" responded his comrade ahead.

And once more the writhing and jingling among the withered leaves. The surveyor strode on, turning his face neither to the right nor to the left, with his Jacob's-staff held upright before him. The other men trooped along scatteringly, dodging under the low boughs of the stunted trees. They pressed hastily together when the great square rocks—Moses' tables of the Law—came into view, lying where it was said the man of God flung them upon the sere slope below, both splintered and fissured, and one broken in twain. The surveyor was bearing straight down upon them. The men running on either side could not determine whether the line would fall within the spot or just beyond. They broke into wild exclamations.

"Ye may hammer me out ez flat ez a skene," cried the blacksmith, "ef I don't b'lieve ez Purdee hev got 'em."

"Naw, sir, naw!" cried another fervent amateur; "thar's the north. I jes now viewed Grinnell's dad's deed; the line ondertakes ter run with Purdee's line; he hev got seven hunderd poles ter the north; ef they air a-goin' ter the north, them tables o' the Law air Grinnell's."

A wild chorus ensued.

"Naw!" "Yes!" "Thar they go!" "A-bearin' off that-a-way!" "Beats my time!" as they stumbled and scuttled alongside the acolytes of the Compass, who bowed down and rose up at every length of the chain. Suddenly a cry from the chain-bearers.

"Out!"

Stillness ensued.

The surveyor stopped to register the "out." It was a moment of thrilling suspense; the rocks lay only a few chains further; Grinnell, into whose confidence doubt had begun to be instilled, said to himself, all a-tremble, that he would hardly have staked his veracity, his standing with the brethren, if he had realized that it was so close a matter as this. He had long known that his father owned the greater part of the unproductive wilderness lying between the two ravines; the land was almost worthless by reason of the steep slants which rendered it utterly untillable. He was sure that by the terms of his deed, which his father had from its vendor, Squire Bates, his line included the Moses' tables on which Purdee had built so fallacious a repute of holiness. He looked once more at the paper—"thence from Crystal Spring with Purdee's line north seven hundred poles to a stake in the middle of the river."

Purdee too was all a-quiver with eagerness. He had not beheld those rocks since that terrible day when all the fine values of his gifted vision had been withdrawn from him, and he could read no more with eyes blinded by the limitations of what other men could see—the infinitely petty purlieus of the average sense. He had a vague idea that should they say this was his land where those strange rocks lay, he would see again, he would read undreamed-of words, writ with a .pen of fire. He started toward them, and then with a conscious effort he held back.

The surveyor took no heed of the sentiments involved in processioning Purdee's land. He stood leaning on his Jacob's-staff, as interesting to him as Moses' rocks, and in his view infinitely more useful, and wiped his brow, and looked about, and yawned. To him it was merely the surveying for a foolish cause of a very impracticable and steep tract of land, and the only reason it should be countenanced by heaven or earth was the fees involved. And this was what he saw at the end of Purdee's line.

Suddenly he took up his Jacob's-staff and marched on with a long stride, bearing straight down upon the rocks. The whole *cortège* started anew— the genuflecting chain-bearers, the dodging, scrambling, running spectators. On one of the strange stunted leafless trees a colony of vagrant crows had perched, eerie enough to seem the denizens of those weird forests; they broke into raucous laughter—Haw! haw! haw!—rising to a wild commotion of harsh, derisive discord as the men once more gave vent to loud, excited cries. For the surveyor, stalking ahead, had passed beyond the great tables of the Law; the chain-bearers were drawing Purdee's line on the other side of them, and they had fallen, if ever they fell here from Moses' hand and broke in twain, upon Purdee's land, granted to his ancestor by the State of Tennessee.

He could not speak for joy, for pride. His dark
eyes were illumined by a glancing, amber light.
He took off his hat and smoothed with his rough
hand his long black hair, falling from his massive
forehead. He leaned against one of the stunted
oaks, shouldering his rifle that he had loaded for
Grinnell—he could hardly believe this, although he
remembered it. He did not want to shoot Grinnell; he would·not waste the good lead !

And indeed Grinnell had much ado to defend
himself against the sneers and rebukes with which
the party beguiled the way through the wintry
woods. "Ter go a-claimin' another man's land,
an' put him ter the expense o' processionin' it, an'
git his line run !" exclaimed the blacksmith, indignantly. "An' ye 'ain't got nare sign o' a show at
Moses' tables !"

"I dunno how this hyar line air a-runnin'," declared Grinnell, sorely beset. "I don't b'lieve it
air a-runnin' north."

The surveyor was hard by. He had planted his
staff again, and was once more taking his bearings.
He looked up for a second.

"North*west*," he said.

Grinnell stared for a moment; then strode up to
the surveyor, and pointed with his stubby finger at
a word on his deed.

The official looked with interest at it; he held
up suddenly Purdee's grant and read aloud, "From
Crystal Spring seven hundred poles north*west* to a
stake in the middle of the river."

He examined, too, the original plat of survey
which he had taken to guide him, and also the plat

made when Squire Bates sold to Grinnell's father; "north*west*" they all agreed. There was evidently a clerical error on the part of the scrivener who had written Grinnell's deed.

In a moment the harassed man saw that through the processioning of Purdee's land he had lost heavily in the extent of his supposed possessions. He it was who had claimed what was rightfully another's. And because of the charge Purdee was the richer by a huge slice of mountain land—how large he could not say, as he ruefully followed the line of survey.

But for this discovery the interest of procession-ing Purdee's land would have subsided with the determination of the ownership of the limited envi-ronment of the stone tables of the Law. Now, as they followed the ever-diverging line to the north-west, the group was pervaded by a subdued and tremulous excitement, in which even the surveyor shared. Two or three whispered apart now and then, and Grinnell, struggling to suppress his dis-may, was keenly conscious of the glances that sought him again and again in the effort to judge how he was taking it. Only Purdee himself was withdrawn from the interest that swayed them all. He had loitered at first, dallying with a temptation to slip silently from the party and retrace his way to the tables and ascertain, perchance, if some vestige of that mystic scripture might not reveal itself to him anew, or if it had been only some morbid fancy, some futile influence of solitude, some fevered condition of the blood or the brain, that had traced on the stone those gracious words, the mere echo of which

—his stuttered, vague recollections—had roused the camp-meeting to fervid enthusiasms undreamed of before. And then he put from him the project —some other time, perhaps, for doubts lurked in his heart, hesitation chilled his resolve—some other time, when his companions and their prosaic influence were all far away. He was roused abruptly, as he stalked along, to the perception of the deepening excitement among them. They had emerged from the dense growths of the mountain to the lower slope, where pastures and fields— whence the grain had been harvested — and a garden and a dwelling, with barns and fences, lay before them all. And as Purdee stopped and stared, the realization of a certain significant fact struck him so suddenly that it seemed to take his breath away. That divergent line stretching to the northwest had left within his boundaries the land on which his enemy had built his home.

He looked; then he smote his thigh and laughed aloud.

The rocks on the river-bank caught the sound, and echoed it again and again, till the air seemed full of derisive voices. Under their stings of jeering clamor, and under the anguish of the calamity which his reeling senses could scarcely measure, Job Grinnell's composure suddenly gave way. He threw up his arms and called upon Heaven; he turned and glared furiously at his enemy. Then, as Purdee's laughter still jarred the air, he drew a "shooting-iron" from his pocket. The blacksmith closed with him, struggling to disarm him. The weapon was discharged in the turmoil, the ball glancing away in the first quiver of sunshine that had reached the earth to-day, and falling spent across the river.

Grinnell wrested himself from the restraining grasp, and rushed down the slope to his gate to hide himself from the gaze of the world—his world, that little group. Then remembering that it was no longer his gate, he turned from it in an agony of loathing. And knowing that earth held no shelter for him but the sufferance of another man's roof, he plunged into the leafless woods as if he heavily dragged himself by a power which warred within him with other strong motives, and disappeared among the myriads of holly bushes all aglow with their red berries.

The spectators still followed the surveyor and his Jacob's-staff, but Purdee lingered. He walked around the fence with a fierce, gloating eye, a panther-like, loping tread, as a beast might patrol a fold before he plunders it. All the venom of the old feud had risen to the opportunity. Here was his enemy at his mercy. He knew that it was less than seven years since the enclosures had been made, acres and acres of tillable land cleared, the houses built—all achieved which converted the worthlessness of a wilderness into the sterling values of a farm. He—he, Roger Purdee—was a rich man for the "mountings," joining his little to this competence. All the cruelties, all the insults, all the traditions of the old vendetta came thronging into his mind, as distinctly presented as if they were a series of hideous pictures; for he was not used to think in detail, but in the full portrayal of scenes.

The Purdee wrongs were all avenged. This result
was so complete, so baffling, so ruinous temporally,
so humiliating spiritually! It was the fullest re-
plication of revenge for all that had challenged it.

"How Uncle Ezra would hev rej'iced ter hev
lived ter see this day!" he thought, with a pious regret that the dead might not know.

The next moment his attention was suddenly attracted by a movement in the door-yard. A woman
had been hanging out clothes to dry, and she turned
to go in, without seeing the striding figure patrol-
ling the enclosure. A baby—a small bundle of a
red dress—was seated on the pile of sorghum-cane
where the mill had worked in the autumn; the stalks
were broken, and flimsy with frost and decay, and
washed by the rains to a pallid hue, yet more
marked in contrast with the brown ground. The
baby's dress made a bright bit of color amidst the
dreary tones. As Purdee caught sight of it he remembered that this was "Grinnell's old baby," who
had been the cause of the renewal of the ancient
quarrel, which had resulted so benignantly for him.
"I owe you a good turn, sis," he murmured, satirically, glaring at the child as the unconscious mother
lifted her to go in the house. The baby, looking
over the maternal shoulder, encountered the stern
eyes staring at her. She stared gravely too. Then
with a bounce and a gurgle she beamed upon him
from out the retirement of her flapping sun-bonnet;
she smiled radiantly, and finally laughed outright,
and waved her hands and again bounced beguil-
ingly, and thus toothlessly coquetting, disappeared
within the door.

Before Purdee reached home, flakes of snow, the first of the season, were whirling through the gray dusk noiselessly, ceaselessly, always falling, yet never seeming to fall, rather to restlessly pervade the air with a vacillating alienation from all the laws of gravitation. Elusive fascinations of thought were liberated with the shining crystalline aerial pulsation; some mysterious attraction dwelt down long vistas amongst the bare trees; their fine fibrous grace of branch and twig was accented by the snow, which lay upon them with exquisite lightness, despite the aggregated bulk, not the densely packed effect which the boughs would show to-morrow. The crags were crowned; their grim faces looked frowningly out like a warrior's from beneath a wreath. Nowhere could the brown ground be seen; already the pine boughs bent, the needles failing to pierce the drifts. On the banks of the stream, on the slopes of the mountain, in wildest jungles, in the niches and crevices of bare cliffs, the holly-berries glowed red in the midst of the ever-green snow-laden leaves and ice-barbed twigs. When his house at last came into view, the roof was deeply covered; the dizzying whirl had followed every line of the rail-fence; scurrying away along the furthest zigzags there was a vanishing glimpse of a squirrel; the boles of the trees were embedded in drifts; the chickens had gone to roost; the sheep were huddling in the broad door of the rude stable; he saw their heads lifted against the dark background within, where the ox was vaguely glimpsed. He caught their mild glance despite the snow that instarred with its ever-shifting crystals the dark space of the aperture, and intervened as a veil. They suddenly reminded him of the season — that it was Christmas Eve; of the sheep which so many years ago beheld the angel of the Lord and the glory of the great light that shone about the shepherds abiding in the fields. Did they follow, he wondered, the shepherds who went to seek for Christ? Ah, as he paused meditatively beside the rail-fence — what matter how long ago it was, how far away! — he saw those sheep lying about the fields under the vast midnight sky. They lift their sleepy heads. Dawn? not yet, surely; and they lay them down again. And one must bleat aloud, turning to see the quickening sky; and one, woolly, white, white as snow, with eyes illumined by the heralding heavens, struggles to its feet, and another, and the flock is astir; and the shepherds, drowsing doubtless, are awakened to good tidings of great joy.

What a night that was! — this night — Christmas Eve. He wondered he had not thought of it before. And the light still shines, and the angel waits, and the eternal hosts proclaim peace on earth, good-will toward men, and summon us all to go and follow the shepherds and see — what? A little child cradled in a manger. The mountaineer, leaning on his gun by the rail-fence, looked through the driving snow with the lights of divination kindling in his eyes, seeing it all, feeling its meaning as never before. Christ came thus, he knew, for a purpose. He could have come in the chariots of the sun or on the wings of the wind. But He was cradled as a little child, that men

might revere humanity for the sake of Him who
had graced it; that they, thinking on Him, might
be good to one another and to all little children.

As he burst into the door of his house the ela-
tions of his high religious mood were rudely dis-
pelled by shrill cries of congratulation from his wife
and her mother. For the news had preceded him.
Ephraim Blinks with his fiddle had stopped there
on his way to play at some neighboring merry-mak-
ing, and had acquainted them with the result of
processioning Purdee's land.

"We'll go down thar an' live!" cried his wife,
with a gush of joyful tears. "Arter all our scratch-
in' along like ten-toed chickens all this time, we'll
hev comfort an' plenty! We'll live in Grinnell's
good house! But ter think o' our trials, an' how
pore we hev been!"

"This air the Purdees' day!" cried the grandmother, her face flushed with the semblance of
youth. "Arter all ez hev kem an' gone, the jedgmint o' the Lord hev descended on Grinnell, an'
he air cast out. An' his fields, an' house, an' bin,
an' barn, air Purdee's!"

The fire flared and faded; shadows of the night
gloomed thick in the room — this night of nights
that bestowed so much, that imposed so much on
man and on his fellow-man!

"Ain't the Grinnell baby got *no* home?" whim-
pered the hereditary enemy.

The mountaineer remembered the Lord of heaven
and earth cradled, a little Child, in the manger.
He remembered, too, the humble child smiling its

guileless good-will at the fence. He broke out
suddenly.

"How kem the fields Purdee's," he cried, leaning his back against the door and striking the
puncheon floor with the butt of the gun till it rang
again and again, "or the house, or the bin, or the
barn? Did he plant 'em? Did he build 'em?
Who made 'em his'n?"

"The law!" exclaimed both women in a breath.

"Thar ain't no law in heaven or yearth ez kin
gin an honest man what ain't his'n by rights," he
declared.

An insistent feminine clamor arose, protesting
the sovereign power of the law. He quaked for a
moment; dominant though he was in his own
house, he could not face them, but he could flee.
He suddenly stepped out of the door, and when
they opened it and looked after him in the snowy
dusk and the whitened woods, he was gone.

And popular opinion coincided with them when
it became known that he had formally relinquished
his right to that portion of the land improved by
Grinnell. He said to the old squire who drew up
the quit-claim deed, which he executed that Christmas Eve, that he was not willing to profit by his
enemy's mistake, and thus the consideration expressed in the conveyance was the value of the
land, considered not as a farm, but as so many
acres of wilderness before an axe was laid to the
trunk of a tree or the soil upturned by a plough.
It was the minimum of value, and Grinnell came
cheaply off.

The blacksmith, the mountain fiddler, and the

advanced thinker, who had been active in the survey, balked of the expected excitement attendant
upon the ousting of Grinnell, and some sensational
culmination of the ancient feud, were not in sympathy with the pacific result, and spoke as if they
had given themselves to unrequited labors.

"Thar ain't no way o' settlin' what that thar
critter Purdee owns 'ceptin' ez consarns Moses'
tables o' the Law.   He clings ter them," they said,
in conclave about the forge fire when the big
doors were closed and the snow, banking up the
crevices, kept out the wind.   "There ain't no use
in percessionin' Purdee's land."

And indeed Purdee's possessions were wider far
than even that divergent line which the county
surveyor ran out might seem to warrant; for on
the mountain-tops largest realms of solemn thought
were open to him.   He levied tribute upon the
liberties of an enthused imagination.   He exulted
in the freedom of the expanding spaces of a spiritual perception of the spiritual things.   When the
snow slipped away from the tables of the Law, the
man who had read strange scripture engraven
thereon took his way one day, doubtful, but faltering with hope, up and up to the vast dome of the
mountain, and knelt beside the rocks to see if perchance he might trace anew those mystic runes
which he once had some fine instinct to decipher.
And as he pondered long he found, or thought he
found, here a familiar character, and there a slowly
developing word, and anon—did he see it aright?—
a phrase; and suddenly it was discovered to him
that, whether their origin were a sacred mystery or

the fantastic scroll-work of time as the rock weathered, high thoughts, evoking thrilling emotions, bear
scant import to one who apprehends only in mental
acceptance. And he realized that the multiform
texts which he had read in the fine and curious
script were but paraphrases of the simple mandate
to be good to one another for the sake of that holy
Child cradled in manger, and to all little children.

THE END

www.ingramcontent.com/pod-product-compliance
Lightning Source LLC
Chambersburg PA
CBHW021357210326
41599CB00011B/908